Topology—Concept of Sets and Spaces

Topology—Concept of Sets and Spaces

N.K. Paswan
Rajesh Bajaj

CYBER TECH PUBLICATIONS
4264/3, Ansari Road, Daryaganj, New Delhi-110002 (India)
Ph.: 011-23244078, 011-43559448 Fax: 011-23280028
E-Mail: cyberpublicationsdelhi@yahoo.com
Website: www.cybertechpublications.com

TOPOLOGY—CONCEPT OF SETS AND SPACES

N.K. Paswan, Rajesh Bajaj

First Edition 2011

Published by :

G.S. Rawat for Cyber Tech Publications
4264/3, Ansari Road, Daryaganj, New Delhi-110002 (India)
Ph.: 011-23244078, 011-43559448 Fax: 011-23280028
E-Mail: cyberpublicationsdelhi@yahoo.com
Website: www.cybertechpublications.com

PREFACE

This books has been specially written to the requirements of the students of all Indian Universities.

Each chapter of this book contains complete theory and fairly large number of solved examples. Sufficient problems has also been selected from various universities examination papers.

The author shall be gratefully to the readers who point out errors and omissions which inspite of all care might have been there. The author will feel amply rewarded if the book serve the purpose for which it is meant. Suggestions for the improvement of the book are always welcome.

The Subject matter has been discussed in such a simple way that the students will find no difficulty to understand it. The proofs of various theorems and examples has been given with minute details.

—*Author*

Contents

Pointwise Convergence, Some Important Inequalities, Holder's Inequality, Cauchy Schwarz Inequality, Minkowski Inequality, Minkowski's Inequality of Norms, Definition : Hilbertspace , Definition Cantor's Tennary Sets, Definition of Neighbourhood, Solved Examples, Exercises.

1

CONCEPT OF SETS

INTRODUCTION

We regard sets and functions as tools of thought, and our purpose in this chapter is to develop these tools to the point where they are sufficiently powerful to serve our needs through the rest of this book. The study of sets and functions leads two ways. One path goes down, into the abysses of logic, philosophy, and the foundations of mathematics. The other goes up, onto the highlands of mathematics itself, where these concepts are indispensable in almost all of pure mathematics as it is today. Needless to say, we follow the latter course. As the reader proceeds, he will come to understand that the words set and *functions* are not as simple as they may seem. In a sense, they are simple; but they are potent words, and the quality of simplicity they possess is that which lies on the far side of complexity.

SET AND CONCEPT

It is well known that any attempt of define a set has always led mathematicians to unsur-mountable difficulties. Most of these things are found in groups. We have around us aggregate of points, flock of birds, assembly of men, bunch of grapes, and so on. Although, in literary language we use different words for different types of groups such as aggregate, flock, assembly, bunch etc., but in mathematics, we use the word collection or set for all types of groups. For example, we define the term set as *"a well defined collection of objects."* One may then ask what is meant by collection. If we answer that a collection is an aggregate of objects or things.

What is then an aggregate? Perhaps then we may define that an aggregate is a class of things. What is then a class? Now we may define a class as a collection. In this manner question after question, since our language is finite, we find that after sometime we will have to use some words have already been questioned. The definition thus becomes circular and worthless. Thus, mathematicians realized that there must be some undefined or primitive terms.

Here we start with two undefined terms—"element" and "set". We assume that the word "set" is synonymous with the words "collection", "aggregate", "class" and is comprised of elements. The words "element", "object", "member" are synonymous. The following explanation will make the idea clear as to what a set actually means. For the objects to be included in a set, it is necessary that they must have some quality common to all yet each element should be distinct from the other. Thus, a formal definition may be given as follows :

Definition : *A set or a collection is group of objects having some well defined and common property or properties. The individual object of collection or set is called an element or a member of the set.*

The following are some examples of sets :

1. The collection of all states in the Indian Unions is a set.
2. The collection of good cricket players of India is not a set, since the term "good player" is vague and it is not well defined.
3. Collection of gentle teachers in a school is not a set, while collection of all teachers in a school is a set.
4. The collection of first five prime numbers is a set containing the elements 2, 3, 5, 7, 11.
5. The collection of past presidents of the Indian Union is a set.
6. The collection of cricketers in the world who scored double century in a test-match is a set.

We shall denote, sets by capital alphabets, *e.g.*, A, B, C, X, Y, Z etc., and the elements by the small alphabets, *e.g.*, a, b, c, x, y, z etc.

During our studies we will have frequent interaction with some sets, so we reserve some letters for these sets as listed below :

N : For the set of natural numbers.

Z : For the set of integers.

Z^+: For the set of all positive integers.

Q : For the set of all rational numbers.

Q^+: For the set of all positive rational numbers.

R : For the set of all real numbers.

R^+: For the set of all positive real numbers.

C : For the set of all complex number.

KINDS OF SETS

Empty Set : *A set is said to be empty or null or void set if it has no element and it is denoted by* ϕ.

In Roster method, ϕ is denoted by {}.

Obviously, a set A is an empty set if the statement $x \in A$ is not true for any x. For example following are the null sets :

(i) A set of even number greater than 8 and less than 10.

(ii) Let $A = \{x : x^2 = 4, x \text{ is odd}\}$. Then B is the empty set because $x^2 = 4 \Rightarrow x = \pm 2$ and 2 is not odd.

Singleton Set : *A set consisting of a single element is called a Singleton Set.*

Examples :

(i) A set of positive integers between 4 and 6 will be a singleton set, consisting of one element 5, *i.e.*, {5}.

(ii) $\{\phi\}$ is a set whose only element is a null set, therefore $\{\phi\}$ is singleton.

Finite Set : *A set is finite if it consists of a finite number of different elements, i.e., if in counting the different members of the set the counting process can come to an end. For example, if A is the set of the week days, it is finite as its elements are seven in number.*

Infinite Set : *A set is infinite if it contains infinite number of elements, i.e., it is not possible to count all members of the set. For example, if B = {2, 4, 6, 8, . . .), then B is infinite.*

Cardinal Numbers of a Finite Set : *The number of elements in a finite set A is called cardinal number* or order of the set A and is denoted by n(A).

Equivalent Sets : *Two finite sets A and B are equivalent if their cardinal numbers are same, i.e.,*

$$n(A) = n(B).$$

Equal Sets : *Two sets A and B are said to be equal if every element of A is a member of B, and every element of B is a member of A.*

If sets A and B are equal, we write A = B and $A \neq B$ when A and B are not equal.

If A = {2, 4, 5, 7} and B = {5, 7, 2, 4}. Then A = B, because each element of A is an element of B and vice-versa. Note that the elements of a set may be listed in any order.

Note : It follows from the above definition of equivalent sets that equal sets are equivalent but equivalent sets need not be equal.

COUNTABLE AND UNCOUNTABLE SETS

The subject of this section and the next—*Infinite Cardinal Numbers*—lies at the very foundation of modern mathematics. It is a vital instrument in the day-to-day work of many mathematicians, and we shall make extensive use of it ourselves. This theory, which was created by the German mathematician Cantor, also has great aesthetic appeal, for it begins with ideas of extreme simplicity and develops through natural stages into an elaborate and beautiful structure of thought. In the course of our discussion we shall answer questions which no one before *Cantor's time* thought to ask, and we shall ask a question which no one can answer to this day.

Without further ado, we can say that *cardinal numbers* are those used in counting, such as the positive integers (or natural numbers) 1, 2, 3, ... familiar to us all. But there is much more to the story than this.

The act of counting is undoubtedly one of the oldest of human activities. Men probably learned to count in a crude way at about the same time as they began to develop articulate speech. The earliest men who lived in communities and domesticated animals must have found it necessary to record the number of goats in the village herd by means of a pile of stones or some similar device. If the herd was counted in each night by removing one stone from the pile for each goat accounted for, then stones left over would have indicated strays, and herdsmen would have gone out to search for them. Names for numbers and symbols for them, like our 1, 2, 3, . . . , would have been superfluous. The simple and yet profound idea of a one-to-one correspondence between the stones and the goats would have fully met the needs of the situation.

In a manner of speaking, we ourselves use the infinite set

$$N + \{1, 2, 3, ...\}$$

of all positive integers as a "pile of stones." We carry this set around with us as part of our intellectual equipment. Whenever we want to count a set, say, a stack of dollar bills, we start through the set N and tally off one bill against each positive integer as we come to it. The last number we reach, corresponding to the last bill, is what we call the number of bills in the stack. If this last number happens to be 10, then "10" is our symbol for the number of bills in the stack, as it also is for the number of our fingers, and for the number of our toes, and for the number of elements in any set, which can be put into one-to-one correspondence with the finite set {1, 2, ... , 10}. Our procedure is slightly more sophisticated than that of the primitive savage. We have the symbols 1, 2, 3, ... for the numbers which arise in counting; we can record them for future use, and communicate them to other people, and manipulate them' by the operations of arithmetic. But the underlying idea, that of the one-to-one correspondence, remains the same for us as it probably was for him.

The positive integers are adequate for the purpose of counting any non-empty finite set, and since outside of mathematics all sets appear to be of this kind, they suffice for all non-mathematical counting. But in the World of Mathematics we are obliged to consider many infinite sets, such as the set of all positive integers itself, the set of all integers, the set of all rational numbers, the set of all real numbers, the set of all points in a plane, and so on. It is often important to be able to count such sets, and it was Cantor's idea to do this, and to develop a thcory of infinite cardinal numbers, by means of one-to-one correspondences.

In comparing the sizes of two sets, the basic concept is that of numerical equivalence as defined in the previous section. We recall that two non-empty sets X and Y are said to be numerically equivalent if there exists a one-to-one mapping of one onto the other, or—and this amounts to the same thing—if there can be found a one-to-one correspondence between them. To say that two non-empty finite sets are numerically equivalent is of course to say that they have the *same number* of *elements* in the ordinary sense. If we count one of them, we simply establish a one-to-one correspondence between its elements and a set of positive integers of the form (1, 2, ... , n}, and we then say that n is the *number of elements possessed by both*, or the *cardinal number of both*. The positive integers are the *finite cardinal numbers*. We encounter many surprises as we follow Cantor and consider numerical equivalence for infinite sets.

The set N = {1, 2, 3, ...} of all positive integers is obviously "larger" than the set (2, 4, 6, ...} of all even positive integers, for it contains this set as a proper subset. It appears on the surface that N has "more" elements. But it is very important to avoid jumping to conclusions when dealing with infinite sets, and we must remember that our criterion in these matters is whether there exists a one-to-one correspondence between the sets (not whether one set is or is not a proper subset of the other). As a matter of fact, the pairing

1, 2, 3, ..., n, ...

2, 4, 6, ..., 2n, ...

serves to establish a one-to-one correspondence between these sets, in which each positive integer in the upper row is matched with the even positive integer (its double) directly below it, and these two sets must therefore be regarded as having the *same number of elements*. This is a very remarkable circumstance, for it seems to contradict our intuition and yet is based only on solid common sense. Every infinite set is numerically equivalent to a proper subset of itself. Since this property is clearly not possessed by any finite set, some writers even use it as the definition of an infinite set.

In much the same way as above, we can show that N is numerically equivalent to the set of *all* even integers;

1, 2, 3, 4, 5, 6, 7, ...

$$0, 2, -2, 4, -4, 6, -6, \ldots$$

Here our device is to start with 0 and follow each even positive integer as we come to it by its negative. Similarly, N is numerically equivalent to the set of all integers:

$$1, 2, 3, 4, 5, 6, 7, \ldots$$
$$0, 1, -1, 2, -2, 3, -3, \ldots$$

It is of considerable historical interest to note that Galileo observed in the early seventeenth century that there are precisely as many perfect squares (1, 4, 9, 16, 25, etc.) among the positive integers as there are positive integers altogether. This is clear from the pairing

$$1, 2, 3, 4, 5, \ldots$$
$$1^2, 2^2, 3^2, 4^2, 5^2, \ldots$$

It struck him as very strange that this should be true, considering how sparsely strewn the squares are among all the positive integers. But the time appears not to have been ripe for the exploration of this phenomenon, or perhaps he had other things on his mind; in any case, he did not follow up his idea.

These examples should make it clear that all that is really necessary in showing that an infinite set X is numerically equivalent to N is that we be able to list the elements of X, with a first, a second, a third, and so on, in such a way that it is completely exhausted by this counting off of its elements. It is for this reason that any infinite set which is numerically equivalent to N is said to be *countably infinite*. We say that a set is *countable* if it is non-empty and finite (in which case it can obviously be counted) or if it is countably infinite.

One of Cantor's earliest discoveries in his study of infinite sets was that the set of all positive rational numbers (which is very large: it contains N and a great many other numbers besides) is actually countable. We cannot list the positive rational numbers in order of size, as we can the positive integers, beginning with the smallest, then the next smallest, and so on, for there is no smallest, and between any two there are infinitely many others. We must find some other way of counting them, and following Cantor, we arrange them not in order of size, but according to the size of the sum of the numerator and denominator. We begin with all positive rationals whose numerator and denominator add up to 2: there is only one, $\frac{1}{1} = 1$. Next we list (with increasing numerators) all those for which this sum is 3: $\frac{1}{2}, \frac{2}{1} = 2$. Next, all those for which this sum is 4: $\frac{1}{3}$, $\frac{2}{2} = 1$, $\frac{3}{1} = 3$. Next, all those for which this sum is 5: $\frac{1}{4}, \frac{2}{3}, \frac{3}{2}, \frac{4}{1} = 4$. Next, all those for which this sum is 6: $\frac{1}{5}$, $\frac{2}{4} = \frac{1}{2}$, $\frac{3}{3} = 1$, $\frac{4}{2} = 2$, $\frac{5}{1} = 5$. And so on. If we now list all these together from the beginning, omitting those already listed when we come to them, we get a sequence

$$1, \frac{1}{2}, 2\frac{1}{3}, 3, \frac{1}{4}, \frac{2}{3}, \frac{3}{2}, 4, \frac{1}{5}, 5 \ldots$$

which contains each positive rational number once and only once. The first row contains all positive rationals with numerator 1, the second all with numerator 2, etc.; and the first column contains all with denominator 1, the second all with denominator 2, and so on. Our listing amounts to traversing this array of numbers as the arrows indicate, where of course all those numbers already encountered are left out. All the infinite sets we considered in the previous section were countable, so it might appear at this stage that *every* infinite set is countable. If this were true, if the end result of the analysis of infinite sets were that they are all numerically equivalent to one another, then Cantor's theory would be relatively trivial. But this is not the case, for Cantor discovered that the infinite set R of all real numbers is *not* countable—or, as we phrase it, R is *uncountable or uncountably infinite*. Since we customarily identify the elements of R with the points of the real line, this amounts to the assertion that the set of all points on the real line represents a "higher type of infinity" than that of only the integral points or only the rational points.

Cantor's proof of this is very ingenious, but it is actually quite simple. In outline the procedure is as follows: we assume that all the real numbers (in decimal form) can be listed, and in fact have been listed; then we produce a real number which cannot be in this list—thus contradicting our initial assumption that a complete listing is possible. In representing real numbers by decimals, we use the scheme of decimal expansion in which infinite chains of 9's are avoided; for instance, we write 1/2 as .5000 . . . and not as .4999 .... In this way we guarantee that each real number has one and only one decimal representation. Suppose now that we can list all the real numbers, and that they have been listed in a column like the one below (where we use particular numbers for the purpose of illustration).

1st number	13 + .712983 . . .
2nd number	–4 + .913572 . . .
3rd number	0 + .843265 ...

Since it is impossible actually to write down this infinite list of decimals, our assumption that all the real numbers can be listed in this way means that we assume that we have available some general rule according to which the list is constructed, similar to that used for listing the positive rationals, and that every conceivable real number occurs somewhere in this list. We now demonstrate that this assumption is false by exhibiting a decimal $.a_1\ a_2\ a_3...$ which is constructed in such a way that it is not in the list. We choose a_1 to be 1 unless the first digit after the decimal point of the first number in our list is 1, in which case we choose a_1 to be 2. Clearly, our new decimal will differ from the first number in our list regardless of how we choose its remaining digits. Next, we choose a_2 to be 1 unless the second digit after the decimal point of the second number in our list is 1, in which case we choose a_2 to be 2. Just as above, our new decimal will necessarily differ from the second number in our list. We continue building up the decimal $.a_1\ a_2\ a_3...$ in this way, and since the process can be continued indefinitely, it defines a real number in decimal form, which is different from each number in our list. This contradicts our assumption that we can list all the real numbers and completes our proof of the fact that the set R of all real numbers is uncountable. The set of all rational points on the real line is countable, and we have just proved that the set of *all* points on the real line is uncountable. We conclude at once from this that irrational points on the real line (*i.e.*, irrational numbers) must exist. In fact, it is very easy to the set of all irrational numbers is uncountably infinite. To vary slightly a striking metaphor coined by E. T. Bell, the rational numbers are spotted along the real line like stars against a black sky, and the dense blackness of the background is the firmament of the irrationals.

The reader is probably familiar with a proof of the fact that the square root of 2 is irrational. This proof demonstrates the existence of irrational numbers by exhibiting a specimen. Our remarks, on the other hand, do not show that this or that particular number is irrational; they merely show that such numbers must exist, and moreover must exist in overwhelming abundance.

One-One Correspondence : If each element of a set A corresponds to one and only one element of another set B and *vice-versa* then we say that there is one-one correspondence between the elements of A and B. The symbol $\leftrightarrow$ is used to express one-to-one correspondence between the elements of two sets. For example, if A = {a, b, c} and B = {book, pen, pencil} then one-one correspondence is shown as given below :

a $\leftrightarrow$ book

b $\leftrightarrow$ pen

c $\leftrightarrow$ pencil.

Remarks:

(1) If the elements of two sets A and B can be put in one-one correspondence they are called *cardinally equivalent.*

(2) Showing one-to-one correspondence between elements, of two sets means pairing of the elements of two sets such that no element of either set is paired with more than one element in the other.

Countable and Uncountable Sets : A set whose elements can be put in one-one correspondence with the elements of the set N, the set of positive integers is called a *countable* or *enumerable set,* otherwise it is *uncountable.*

Comparability of Sets : Two sets A and B are said to be *comparable* if $A \subset B$ or $B \subset A$. If neither $A \subset B$ nor $B \subset A$ then A and B are called *non-comparable.* For example,

(i) $A = \{2, 4\}$ and $B = \{2, 4, 5, 6\}$ are comparable as $A \subset B$.

(ii) If $A = \{x : x \text{ is an even integer}\}$

and $B = \{ x : x \text{ is an odd integer}\}$

then A and B are non-comparable because $A \not\subset B$ and $B \not\subset A$.

UNIVERSAL SET

In any discussion in set theory, there always happens to be a set that contains all set under consideration, *i.e.*, it is a super set of each of the given sets. Such a set is called the universal set and is denoted by U.

Definition : *A set that contains all sets in a given context is called the Universal Set.*

Examples :

(i) If $A = \{1, 2\}$, B $\{2, 4, 5\}$ and $C = \{1, 3, 5, 7\}$, then $U = \{1, 2, 3, 4, 5, 7)$ can be written as universal set.

(ii) When we study a set of even numbers, set of odd numbers and the set of prime numbers then the set of natural numbers is universal set.

(iii) When we refer to two dimensional coordinate geometry, then the set of all points in xy-plane is the universal set.

SETS AND ITS INCLUSION

We adopt a naive point of view in our discussion of sets and assume that the concepts of an element and of a set of elements are intuitively clear. By an *element* we mean an object or entity of some sort, as, for example, a positive integer, a point on the real line (= a real number), or a point in the complex plane (= a complex number). A *set* is a, collection or aggregate of such elements, considered together or as a whole. Some examples are furnished by the set of all even positive integers, the set of all rational points on the real line, and the set of all points in the complex plane whose distance from the origin is 1 (= the unit circle in the plane). We reserve the word *class* to refer to a set of sets. We might speak, for instance, of the class of all circles in a plane (thinking of each circle as a set of points). It will be useful in the work we do if we carry this hierarchy one step further and use the term *family* for a set of classes. One more remark: the words element, set, class, and family are not intended to be rigidly fixed in their usage; we use them fluidly, to express, varying attitudes toward the mathematical objects and systems we study.

It is entirely reasonable, for instance, to think of a circle not as a set of points, but as a single entity in itself, in which case we might justifiably speak of the set of all circles in a plane. There are two standard notations available for designating a particular set. Whenever it is feasible to do so, we can list its elements between braces. Thus {l, 2, 3} signifies the set consisting of the first three positive integers, {1, i, –1, –i} is the set of the four fourth roots of unity, and {–1, –3, –5, . . ., i} is the set of all odd integers.

This manner of specifying a set, by listing its elements, is unworkable in many circumstances. We are then obliged to fall back on the second method, which is to use a property or attribute that characterizes the elements of the set in question. If P denotes a certain property of elements, then {x : P} stands for the set of all elements x for which the property P is meaningful and true. For example, the expression,

{x : x is real and irrational},

which we read the *set of all x such that x is real and irrational*, denotes the set of all real numbers which cannot be written as the quotient of two integers. The set under discussion contains all those elements (and no others) which possess the stated property. The three sets of numbers described at the beginning of this paragraph can be written either way:

$\{1, 2, 3\} = \{n : n \text{ is an integer and } 0 < n < 4\}$,

$\{1, i, -1, -i\} = \{2 : 2 \text{ is a complex number and } z^4 = 1\}$,

and $\{\pm l, -3, -5, \ldots\} = \{n : n \text{ is an odd integer}\}$.

We often shorten our notation. For instance, the last two sets mentioned might perfectly well be written $\{z : z^4 = 1\}$ and $\{n : n \text{ is odd}\}$. Our purpose is to be clear and to avoid mis-understandings, and if this can be achieved with less notation, so much the better. In the same vein we can write,

the unit circle $= \{z : |z| = 1\}$,

the closed unit disc $= \{z : |z| \leq 1\}$,

and the open unit disc = $\{z : |z| < 1\}$,

We use a special system of notation for designating intervals of various kinds on the real line. If a and b are real numbers such that a < b, then the following symbols on the left are defined to be the indicated sets on the right:

$$[a,b] = \{x : a < x < b\},$$

$$(a,b] = \{x:a < x < b\},$$

$$[a,b) = \{x:a <, x < b\},$$

$$(a,b) = \{x:a < x < b\}.$$

We speak of these as the closed, the open-closed, the closed-open, and the open intervals from a to b. In particular, [0, 1] is the *closed unit* interval, and (0, 1) is the *open unit interval.*

There are certain logical difficulties which arise in the foundations of the theory of sets. We avoid these difficulties by assuming that each discussion in which a number of sets are involved takes place in the context of a single fixed set. This set is called the *universal set*. It is denoted by U, every set mentioned is assumed to consist of elements in U.

There will always be on hand a given space within which we work, and this will serve without further comment as our universal set. It is often convenient to have available in U a set containing no elements whatever; we call this the *empty set* and denote it by the symbol ϕ. A set is said to be *finite* if it is empty or consists of n elements for some positive integer n; otherwise, it is said to be *infinite.*

We usually denote elements by small letters and sets by large letters. If x is an element and A is a set, the statement that x is an element of A (or belongs to A, or is contained in A) is symbolized by $x \varepsilon A$. We denote the negation of this, namely, the statement that x is not an element of A, by $x \notin A$.

Two sets A and B are said to be *equal* if they consist of exactly the same elements; we denote this relation by A = B and its negation by $A \neq B$. We say that A is a *subset* of B (or is contained in B) if each element of A is also an element of B. This relation is symbolized by $A \subseteq B$.

We sometimes express this by saying that B is a *superset* of A (or contains A). $A \subseteq B$ allows for the possibility that A and B might be equal. If A is a subset of B and is not equal to B, we say that A is a *proper subset* of B (or is properly contained in B).

This relation is denoted by $A \subset B$. We can also express $A \subset B$ by saying that B is a *proper superset* of A (or properly contains A). The relation $\subseteq$ is usually called set inclusion. We sometimes reverse the symbols introduced in the previous paragraph. Thus $A \subseteq B$ and $A \subset B$ are occasionally written in the equivalent forms $B \supseteq A$ and $B \supset A$.

It will often be convenient to have a symbol for logical implication, and $\Rightarrow$ is the symbol we use. If p and q are statements, then $p \Rightarrow q$ means that p *implies* q, or that if p is true, then q is also true. Similarly, $\Leftrightarrow$ is our symbol for two-way implication or logical equivalence. It means that the statement on each side implies the statement on the other, and is usually read if *and only* if, or *is equivalent to*.

The main properties of set inclusion are obvious. They are the following:

(1) $A \subseteq A$ for every A;

(2) $A \subseteq B$ and $B \subseteq A \Rightarrow A = B$;

(3) $A \subseteq B$ and $B \subseteq C \Rightarrow A \subseteq C$.

It is quite important to observe that (1) and (2) can be combined into the single statement that $A = B \Leftrightarrow A \subseteq B$ and $B \subseteq A$. This remark contains a useful principle of proof, namely, that the only way to show that two sets are equal, apart from merely inspecting them, is to show that each is a subset of the other.

STATEMENTS

We communicate our ideas to others through sentences. In our mathematical language, *the sentences which can be judged to be true or false are known as statements*. For Example, $2 + 3 = 5$ is a true statement while, $(3 + 4) < 6$ is a false statement. Also, Ram is a handsome boy, is a sentence, not a statement. We use connectives *'and'* and *'or'* to combine two statements to form a compound statement.

We know that for any three numbers a, b, c we always have

$$a\ (a + b) = a.b + a.c.$$

We say that in numbers, multiplication distributes addition.

But $a + (bc) \neq (a + b), (a + c)$ *i.e.*, addition does not distribute multiplication is numbers. However, is statements 'and' *distributes* 'or' and 'or' *distributes* 'and'.

Thus for any statements p, q and r we have,

p and (q or r) = (p & q) or (p & r)

and p or (q & r) = (p or q) and (P or r).

We shall be using the symbols '$\forall$' and '$\exists$' for *'for all values of'* and *'there exists'* respectively. Also,

(i) If ($p \Rightarrow q$ and $q \Rightarrow p$), the we write, '$p \Rightarrow q$' *i.e.,* p *implies and is implied by q.*

(ii) We write, '$p \Rightarrow q$', if whenever p is true, the q is true and we say that, *p implies q.*

SET NOTATIONS

If an element a belongs to a set A, then we express it by writing :

$$a \in A.$$

Which can also be read as "a is an element of the set A" or "A contains a as one of its elements." If on the contrary, an element a is not a member of a set A, then we write

$$a \notin A.$$

A set can be specified in following two ways. We can make use of any of the two notations according our convenience.

(a) Roster Method (Tabular Form) : In this method, a set is represented by listing all its elements within braces {}. For example, the set of vowels in the English alphabets would be written as {a, e, i, o, u}.

(b) Rule Method (Set-builder Form) : In this method, a set is represented by describing its elements in terms of one or several characteristic properties which enable us to decide whether a given object is an element of the set under consideration or not. For example, the set of all natural numbers can be denoted as follows :

N = {x : x is a natural number} or

{x | x is a natural number} which is read as the set of all x such that x belongs to N.

The symbol : or/stands for 'such that' and $\in$ for 'belongs to' as explained earlier. A few examples of set notations by Roster and Rule Methods are given below :

(i) Set of all odd numbers.

(a) O = {1, 3, 5, . . .) (Roster method)

(b) O = {x : x is an odd number} (Rule method)

(ii) Set of all odd numbers between 2 and 10.

(a) B = {3, 5, 7, 9, . . .) (Roster method)

(b) B = {x : x < 0, 2 < x < 10} (Rule method)

Remark : It is worth mentioning that a set is not just any collection. For a collection to be a set it is essential that given any element, one should be able to decide whether it does or does not belong to the set. The assumption that every collection is a set leads to a paradox known as :

Russel Paradox : The following example will illustrate the basic idea underlying this Paradox.

A barber in a certain town shaved all those and only those, who did not shave themselves. If A be the collection of all those whom the barber shaved, is the barber a member of A or not?

The two answer are possible :

(i) The barber is a member of A.

(ii) The barber is not a member of A.

But these two are contradictory statements for if the barber is a member of A, then the barber has shaved himself and since he has shaved himself, he can not be a member of A.

SUB SETS

If every element in a set A is also a member of a set B, then A is called a subset of B. More specifically, A is a subset of B if $x \in A \Rightarrow x \in B$.

The symbol $\Rightarrow$ stands for "implies". We denote this relationship by writing

$$A \subset B.$$

Which can also be read as "A is contained is B". Also B is called the *Super Set* of A and can be written as $B \supset A$. Thus if $x \in A \Rightarrow x \in B$, then $A \subset B$.

Examples :

(i) Let $A = \{x : x \text{ is even}\}$, *i.e.*, $A = \{2, 4, 6, \ldots)$ and let $B = \{x : x$ is a positive power of $2\}$, *i.e.*, $B = \{2, 4, 8, 16\}$. Then $A \subset B$, *i.e.*, A is contained in B.

(ii) If P be the set of all parallelograms and S is the set of all squares in a plane, *i.e.*,

$P = \{$all parallelograms in the plane$\}$

and $S = \{$all squares in the plane$\}$

Then S is a subset of P, *i.e.*,

$S \subset P$.

Note : Some authors use symbol $\subseteq$ to express a subset. If A is not a subset of B, we write it as $A \not\subset B$ or $A \not\subseteq B$.

Remarks: (1) If we have to prove $A \subset B$, then we should prove that

$$x \in A \Rightarrow x \in B.$$

Symbolically, $A \subset B$ iff $(x \in A \Rightarrow x \in B)$.

(2) If we have to prove that $A \not\subset B$ then we should show that there exists atleast one element x such that

$$x \in A \text{ but } x \notin B.$$

Symbolically, $A \not\subset B$ iff $\{\exists\, x \in A$ s.t. $x \notin B)$.

The symbol $\exists$ stands for 'there exists' and s.t. for 'such that'.

Example : *Write the subset of the following set :*

$$A = \{1, 2, 3\}$$

Solution : The required subsets are :

$$\phi, \{1\}, \{2\}, \{3\}, \{1, 2\},$$
$$\{1, 3\}, \{2, 3\}, \{1, 2, 3\}.$$

Equality of Sets : Set A is said to be equal to set B if both of them have the same members, *i.e.*, every element which belongs to A also belongs to B ($A \subset B$) and every element which belongs to B also belongs to A ($B \subset A$). We denote the equality of sets A and B by

$$\Rightarrow A = B.$$

Proper Subsets : Let A and B be two sets. If $B \subset A$ and $B \neq A$, then B is said to be a proper subset of A.

For example, if A $\{1, 2, 3\}$ and $B = \{2, 3\}$ then B is proper subset of A because all the elements of B are in A but one element 1 of A is not in B.

Theorem 1:

Null set ϕ is a subset of every set.

Proof :

Let A be any given set. We shall prove that $\phi \subset A$. Let us suppose, on contrary, that $\phi \not\subset A$.

$\phi \not\subset A \Rightarrow$ there is atleast one element x such that $x \in \phi$

and $\qquad x \notin A \qquad$ ...(1)

But since ϕ is a null set,

therefore $\qquad x \notin \phi \qquad$ ...(2)

Therefore, (1) and (2) $\Rightarrow x \in \phi$ and $x \notin \phi$, which is absurd and therefore, our assumption is wrong. Consequently $\phi \subset A$.

Since A is an arbitrary set ϕ is a subset of every set.

Theorem 2:

Every set is a subset of itself.

Proof :

Let A be any set. Then each element of A is clearly in A itself. Hence $A \subset A$.

Theorem 3:

If $A \subset B$ and $B \subset C$ then $A \subset C$.

Proof :

We must show that every element in A is also an element is C. Let x be an element of A, *i.e.*,

$$x \in A.$$

But it is given that $A \subset B$.

$$\therefore \quad x \in A \Rightarrow x \in B \quad ...(1)$$

Also

$$B \subset C$$

$$\therefore \quad x \in B \Rightarrow x \in C \quad ...(2)$$

Hence from (1) and (2)

$$x \in A \Rightarrow x \in C$$

$$\therefore \quad A \subset C.$$

Number of Subsets of a Finite Set : Consider a set {a}. It has two possible subsets {a} and ϕ. Again for the set {a, b}, the possible subsets are ϕ, {a}, {b}, {a, b} which are 4 subsets in number. A set {a, b, c} has as many as the following subsets.

$$\phi, \{a\}, \{b\}, \{c\}, \{a, b\}, \{b, c\}, \{a, c\}, \{a, b, c\}.$$

which are eight in numbers.

Thus we can generalize the result as follows :

A set with 1 element has 2^1 subsets.

A set with 2 element has 2^2 subsets.

A set with 3 element has 2^3 subsets.

Theorem 4:

The total number of subsets of a finite set containing n elements is 2^n.

Proof:

Let A be a finite set containing n elements. Let $O \leq r \leq n$. Consider those subsets of A that have r elements each. We know that the number of ways in which r elements can be chosen out of n elements is ${}^nC_r$. Therefore, the number of subsets of A having r elements each is ${}^nC_r$.

Hence, the total number of subsets of A.

$$= {}^nC_0 + {}^nC_1 + {}^nC_2 + \ldots + {}^nC_r + \ldots + {}^nC_n$$
$$= (1 + 1)^n = 2^n.$$

POWER SET

The set of all subsets of a given set A is called the power set of A and is denoted by the symbol P(A).

i.e., $P(A) = \{T : T \subset A\}$

ϕ and A are both members of P(A).

Example : Let $A = \{a, b\}$

then $P(A) = \{\phi, \{a\}, \{b\}, \{a, b\}\}$.

VENN DIAGRAMS

Sometimes pictures are very helpful in our thinking. Swiss mathematician Euler, first of all gave an idea to represent a set by the points in a closed curve (usually a circle but not necessarily circle). Later on British Mathematician Venn brought this idea to practice. That is why the diagrams drawn for this purpose are called Venn-Euler diagrams or simply Venn-diagrams. In Venn diagrams the universal set U is represented by points within a rectangle and its subsets are represented by the points in closed curves (usual circles) within the rectangle. These diagrams are very useful for the beginners to understood the set theoric ideas, though they are not of much importance as the records advances.

The idea of $A \subset B$ and $A \neq B$ can be represented by the following diagram.

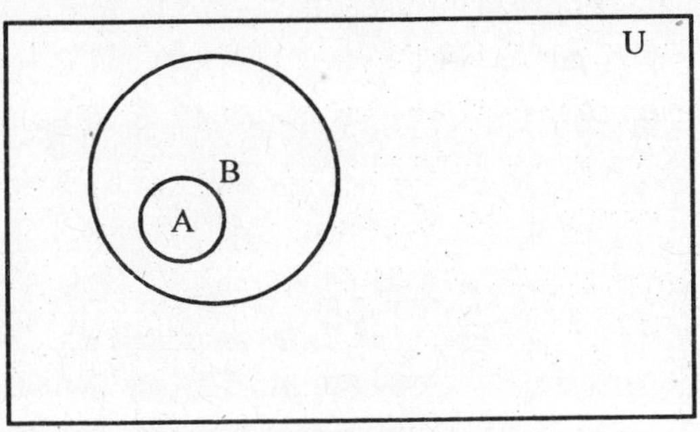

Fig. 1.1

BASIC OPERATIONS ON SETS

Union of Sets : Let A and B be two given sets. *The set which contains every element contained in A or B or both A and B is called the union (or join) of A and B.* In fact union is an 'either' 'or' idea. The symbol ∪ is used to denote the union of sets. Thus, A ∪ B is read as 'A union B' or 'A join B' or 'A Cup B'. Symbolically, $A \cup B = \{x : x \in A \text{ or } x \in B\}$.

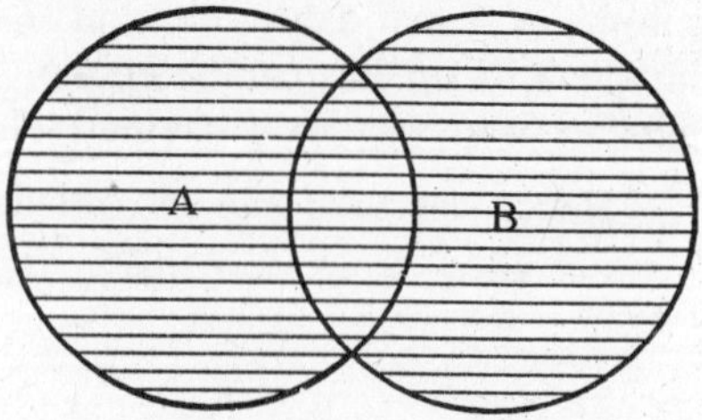

Fig. 1.2 : A ∪ B (Shaded).

The shaded part represent A ∪ B in the following Venn diagram :

Examples :

(i) Let $A = \{1, 2, 3\}$

and $B = \{2, 4, 6\}$,

then $A \cup B = (1, 2, 3, 4, 6\}$.

(ii) If $A = \{x : x = 2n + 1, n \in Z\}$

and $B = \{x : x = 2n, n \in Z\}$.

then $A \cup B = \{x : x \text{ is an odd integer}\}$

$\cup\{x : x \text{ is an even integer}\}$

$= \{x : x \text{ is an integer}\} = Z$.

Notes :

(1) Each element in a set is listed once only because the repetition of elements is meaningless in a set. If A is a set of cricket players, B is a set of tennis players and some players are common to both the teams, then

A ∪ B = Set of all players in the two teams.

(2) From the definition it is clear that A and B are the subsets of A ∪ B.

Symbolically, $A \subset (A \cup B)$ and $B \subset (A \cup B)$.

(3) The union of a finite number of sets $A_1, A_2, \ldots, A_n$ is denoted by

$$A_1 \cup A_2 \cup A_3, \ldots \cup A_n$$

or by $\bigcup_{i=1}^{n} A_1$.

Intersection of Sets : Let A and B be two given sets. *The set containing all the elements which are contained in A as well as in B is called the intersection of A and B.* This is why it is said that intersection is an 'and' idea. The symbol ∩ is used to denote the intersection of sets.

Thus, $A \cap B$ is read as the intersection of A and B.

Symbolically, $A \cap B = \{x : x \in A \text{ and } x \in B\}$.

In the following Venn diagram the shaded part represents $A \cap B$.

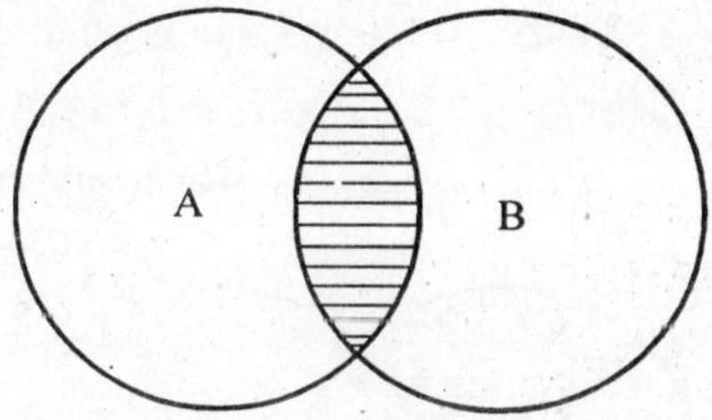

Fig. 1.3 : A ∩ B (Shaded).

Examples :

(i) If $A = \{a, b, i, z\}$

$B = \{b, c, i, u, v\}$

Then $A \cap B = \{b, i\}$.

(ii) If $A = \{x : x = 2n, n \in Z\}$ and

$B = \{x : x = 3n, n \in Z\}$,

Then $A \cap B = \{x : x = 2n, n \in Z\} \cap \{x : x = 3n, n \in Z\}$

$= \{\ldots, -4, -2, 0, 2, 4, 6, \ldots\} \cap \{\ldots, -9, -6, -3, 0, 3, 6, 9, \ldots\}$

$= \{\ldots, -6, 0, 6, 12, \ldots\} = \{x : x = 6n, n \in Z\}$.

Notes :

(1) If A and B are any two sets, then

$A \cap B \subset A$ and $A \cap B \subset B$.

(2) The intersection of finite number of sets

$A_1, A_2, \ldots, A_n$ is denoted by

$A_1 \cap A_2 \cap \ldots \cap A_n$

or by $\bigcap_{i=1}^{n} A_1$

Disjoint Sets : *Two sets A and B are said to be disjoint sets if they have no element in common, i.e., if their intersection is a null set, i.e.,*

$$A \cap B = \phi.$$

Thus, $A \cap B = \phi \Rightarrow$ A and B are disjoint. For example, if

$A = \{a, b, c\}$ and $B = \{p, q, r\}$

then A and B are disjoint sets as there is no element common to both of them.

Difference of Two Sets : Let A and B be two sets. The difference of A and B, written as $A - B$, is the set of all those elements of A which do not belong to B.

Thus, $A - B = \{ x : x \in A \text{ and } x \notin B\}$

or $A - B = \{ x \in A : \text{and } x \notin B\}$.

In Fig. 1.4 (i) the shaded part represents A – B.

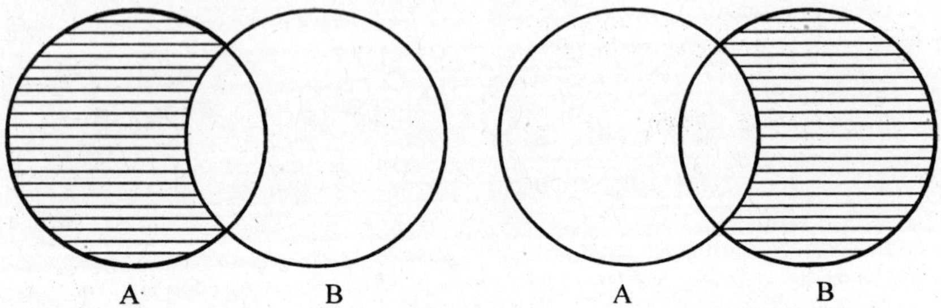

A–B (Shaded)
(i)

B–A (Shaded)
(ii)

Fig. 1.4

Similarly the difference B – A is the set of all those elements of B that do not belong to A, *i.e.*,

$$B - A = \{x \in B : x \notin A\}$$

In Fig. 1.4 (ii) the shaded part represents B – A

Example : If $A = \{2, 3, 4, 5, 6, 7\}$

and $B = \{ 3, 5, 7, 9, 11, 13\}$,

then $A - B = \{2, 4, 6\}$

and $B - A = \{9, 11, 13\}$.

Symmetric Difference of Two Sets : Let A and B be two sets. The symmetric difference of sets A and B is the set $(A - B) \cup (B - A)$ and is denoted by $A \Delta B$.

Thus, $A \Delta B = (A - B) \cup (B - A)$

$= \{x : x A \cap B\}$

In following figure shaded part represents $A \Delta B$.

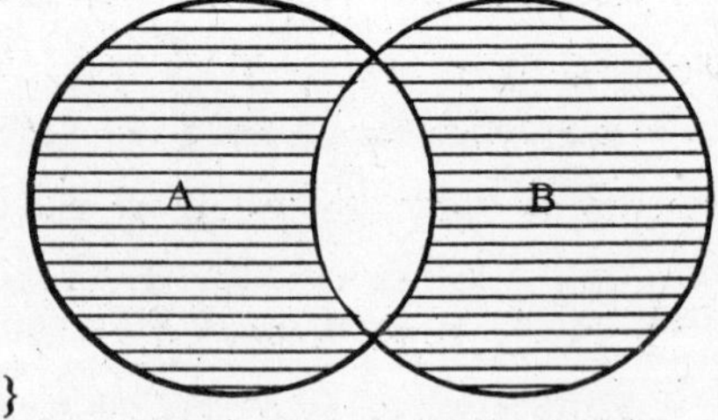

Fig. 1.5 : A Δ B (Shaded).

For example, If $A = \{1, 2, 3\}$

$B = \{3, 4, 5)$

then $A - B = \{1, 2\}$

$B - A = \{4, 5\}$

$\therefore \quad A \Delta B = \{1, 2, 4, 5\}$.

Complement of a Set : The complement of a given set A is defined as the set consisting of those elements of the universal set which are not contained in the given set A. It is denoted by the symbol A' or $\overline{A}$.

Symbolically, $A' = \{x : x \in U, x \notin A\}$.

In the following Venn-diagram shaded portion represents A'.

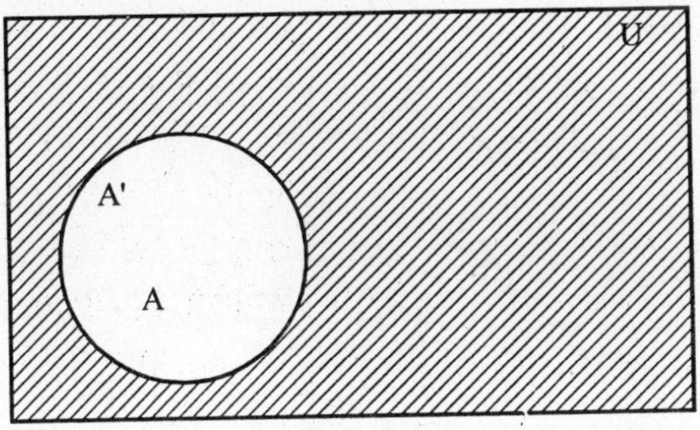

Fig. 1.6

Example : Le N, the set of all natural numbers be taken as the universal set and let

$$A = \{x : x \text{ is an even number and } x \in N\},$$

then $$A' = \{x : x \text{ is an odd number, } x \in N\}.$$

Notes :

(1) The set A and its complement A' are disjoint sets.

(2) Let A and B be two sets. The set $X = \{x : x \in A, x \notin B)$ is called the complement of the set B with respect to A (*i.e.*, A – B).

COUNTING PRINCIPLE

If A, B and C are finite sets, and U be the finite universal set, then

(i) $n(A \cup B) = n(A) + n(B) - n(A \cap B)$

(ii) $n(A \cup B) = n(A) + n(B) \Leftrightarrow$ A, B are disjoint non-void sets

(iii) $n(A - B) = n(A) - n(A \cap B)$, *i.e.*,

$n(A - B) + n(A \cap B) = n(A)$.

(iv) $n(A \Delta B)$ = No. of elements which belong to exactly one of A or B

$= n((A - B) \cup (B - A)$

$= n(A - B) + n(B - A)$ ($\because$ (A–B) and (B–A) and disjoint)

$= n(A) - n(A \cap B) + n(B) - n(A \cap B)$

$= n(A) + n(B) - 2n(A \cap B)$

(v) $n(A \cup B \cup C) = n(A) + n(B) + n(C) - n(A \cap B) - n(B \cap C) - n(A \cap C) + n(A \cap B \cap C)$

(vi) No. of elements in exactly two of the sets A, B, C

$$= n(A \cap B) + n(B \cap C) + n(C \cap A) - 3n(A \cap B \cap C)$$

(vii) No. of elements in exactly one of the sets A, B, C.

$$= n(A) + n(B) + n(C) - 2n(A \cap B) - 2n(B \cap C) - 2n(A \cap C) + 3n(A \cap B \cap C)$$

(viii) $n(A' \cup B') = n((A \cap B)')$

$$= n(U) - n(A \cap B)$$

(ix) $n(A' \cap B') = n((A \cup B)')$

$$= n(U) - n(A \cup B).$$

ORDERED PAIRS

If x and y be any two elements, then (x, y) is called their ordered pair. The element x is said to be the first member or the first coordinate of (x, y) and the element y is called the second member or coordinate of the ordered pair (x, y). It is clear from here that

$$(a, b) = (c, d) \Leftrightarrow a = c, b = d$$

Example : If ordered pairs $(2x - 1, -5)$ and $(x, y + 1)$ are equal, then

$$2x - 1 = x \Rightarrow x = 1$$

and $\quad -5 = y + 1 \Rightarrow y = -6.$

DUALITY PRINCIPLE

It may be noticed that if any law of the algebra of sets, universal set U is replaced by ϕ, ϕ by U, $\cup$ by $\cap$ and $\cap$ by $\cup$ wherever these occur, the new statement thus obtained is also a law of the algebra of sets. This fact is known as the *Duality Principle,* and any law obtained as a result of its application is called the dual of the original law. It is sometimes called primal law. For example, the dual of the law.

$$A \cup (B \cap C) = (A \cup B) \cap (A \cup C)$$

is $\quad A \cap (B \cup C) = (A \cap B) \cup (A \cap C)$

It will be observed that this principle is always true.

LAWS OF ALGEBRA OF SETS

I. Idempotent Laws : *For any set A, we have*

(i) $A \cup A = A$,

(ii) $A \cap A = A$

Proof :

(i) $A \cup A = \{x : x \in A \text{ or } x \in A\}$

$= \{x : x \in A\} = A$

(ii) $A \cap A = \{x : x \in A \text{ and } x \in A\}$

$= \{x : x \in A\} = A.$

II. Identity Laws : *For any set A, we have,*

(i) $A \cup \phi = A$

(ii) $A \cap U = A$

i.e., ϕ and U are identity elements for union and intersection respectively.

Proof :

(i) $A \cup \phi = \{x : x \in A \text{ or } x \in \phi\}$

$= \{x : x \in A\} = A$

(ii) $A \cap U = \{x : x \in A \text{ and } x \in U\}$

$= \{x : x \in A\} = A.$

III. Commulative Laws : *For any sets A and B, union and intersection are commulative, i.e.,*

(i) $A \cup B = B \cup A$, *(ii)* $A \cap B = B \cap A$.

Proof :

(i) Let x be an arbitrary element of $A \cup B$. Then

$x \in A \cup B \Rightarrow x \in A \text{ or } x \in B$

$\Rightarrow x \in B \text{ or } x \in A$

$\Rightarrow x \in B \cup A$

$\therefore A \cup B \subset B \cup A$

Similarly, $B \cup A \subset A \cup B$

$\therefore A \cup B = B \cup A.$

(ii) Let y be an arbitrary element of $A \cap B$. Then

$x \in A \cap B \Rightarrow y \in A \text{ and } y \in B$

$\Rightarrow y \in B \text{ and } y \in A$

$\Rightarrow y \in B \cap A$

$\therefore A \cap B \subset B \cap A$

Similarly, $B \cap A \subset A \cap B$

$\therefore A \cap B = B \cap A.$

IV. Associative Laws : *If A, B and C are any three sets, then*

(i) $(A \cup B) \cup C = A \cup (B \cup C)$

(ii) $A \cap (B \cap C) = (A \cap B) \cap C$

i.e. union and intersection are associative.

Proof :

(i) Let x be an arbitrary element of $(A \cup B) \cup C$. Then

$x \in (A \cup B) \cup C$

$\Rightarrow \quad x \in (A \cup B)$ or $x \in C$

$\Rightarrow \quad (x \in A$ or $x \in B)$ or $x \in C$

$\Rightarrow \quad x \in A$ or $(x \in B$ or $x \in C)$

$\Rightarrow \quad x \in A$ or $(x \in B \cup C)$

$\Rightarrow \quad x \in A \cup (B \cup C)$

$\therefore \quad (A \cup B) \cup C \subset A \cup (B \cup C)$

Similarly, $A \cup (B \cup C) \subset (A \cup B) \cup C$

$\therefore \quad (A \cup B) \cup C = A \cup (B \cup C)$

(ii) Let y be an arbitrary element of $A \cap (B \cap C)$. Then

$y \in A \cap (B \cap C)$

$\Rightarrow \quad y \in A$ and $y \in (B \cap C)$

$\Rightarrow \quad y \in A$ and $(y \in B$ and $y \in C)$

$\Rightarrow \quad (y \in A$ and $y \in B)$ and $y \in C$

$\Rightarrow \quad y \in (A \cap B)$ and $y \in C$

$\Rightarrow \quad y \in (A \cap B) \cap C$

$\therefore \quad A \cap (B \cap C) \subset (A \cap B) \cap C$

Similarly, $(A \cap B) \cap C) \subset A \cap (B \cap C)$

Hence, $\quad A \cap (B \cap C) = (A \cap B) \cap C.$

V. Distributive Laws : *If A, B and C are any three sets, then*

(i) $A \cup (B \cup C) = (A \cup B) \cap (A \cup C)$

(ii) $A \cap (B \cap C) = (A \cap B) \cup (A \cap C)$

i.e. union and intersection are distributive over intersection and intersection respectively.

Proof :

(i) Let x be an arbitrary element of $A \cup (B \cap C)$. Then

$x \in A \cup (B \cap C)$

$\Rightarrow \quad x \in A$ or $x \in (B \cap C)$

$\Rightarrow \quad (x \in A)$ or $(x \in B$ and $x \in C)$

$\Rightarrow \quad (x \in A$ or $x \in B)$ and $(x \in A$ or $x \in C)$

[$\because$ 'or' is distributive over 'and']

$\Rightarrow \quad x \in (A \cup B)$ and $x \in (A \cup C)$

$\Rightarrow \quad x \in (A \cup B) \cap (A \cup C)$

$\therefore \quad A \cup (B \cap C) \subset (A \cup B) \cap (A \cup C)$

Similarly, $(A \cup B) \cap (A \cup C) \subset A \cup (B \cap C)$

Hence, $A \cup (B \cap C) = (A \cup B) \cap (A \cup C)$

(ii) Let y be an arbitrary element of $A \cap (B \cup C)$. Then

$x \in A \cap (B \cup C)$

$\Rightarrow \quad x \in A$ and $x \in (B \cup C)$

$\Rightarrow \quad x \in A$ and $(x \in B$ or $x \in C)$

$\Rightarrow \quad (x \in A$ and $x \in B)$ or $(x \in A$ and $x \in C)$

$\Rightarrow \quad x \in (A \cap B)$ or $x \in (A \cap C)$

$\Rightarrow \quad x \in (A \cap B) \cup (A \cap C)$

$\therefore \quad A \cap (B \cap C) \subset (A \cap B) \cup (A \cap C)$

Similarly, $(A \cap B) \cup (A \cap C) \subset A \cap (B \cup C)$

$\therefore \quad A \cap (B \cup C) = (A \cap B) \cup (A \cap C)$.

VI. De-Morgan's Laws : *If A, B are any two sets, then*

(i) $(A \cup B)' = A' \cap B'$

(ii) $(A \cap B)' = A' \cup B'$

Proof :

(i) Let x be an arbitrary element of $(A \cup B)'$. Then

$x \in (A \cup B)'$

$\Rightarrow \quad x \notin (A \cup B)$

$\Rightarrow \quad x \notin A$ and $x \notin B$

$\Rightarrow \quad x \in A'$ and $x \in B'$

$\Rightarrow \quad x \in A' \cap B'$

$\therefore \quad (A \cup B)' \subset A' \cap B'$

Again, let y be an arbitrary element of $A' \cap B'$. Then

$$
\begin{aligned}
& y \in A' \cap B' \\
\Rightarrow\quad & y \in A' \text{ and } y \in B' \\
\Rightarrow\quad & y \notin A \text{ and } y \notin B \\
\Rightarrow\quad & y \notin A \cup B \\
\Rightarrow\quad & y \notin (A \cup B)' \\
\therefore\quad & A' \cap B' \subset (A \cup B)'
\end{aligned}
$$

Hence, $(A \cup B)' = A' \cap B'$.

(ii) Let x be an arbitrary element of $(A \cap B)'$. Then

$$
\begin{aligned}
& x \in (A \cap B)' \\
\Rightarrow\quad & x \notin (A \cap B) \\
\Rightarrow\quad & x \notin A \text{ or } x \notin B \\
\Rightarrow\quad & x \in A' \text{ or } x \in B' \\
\Rightarrow\quad & x \in A' \cup B' \\
\Rightarrow\quad & (A \cap B)' \subset A' \cup B'
\end{aligned}
$$

Again, let y be an arbitrary element of $A' \cup B'$. Then

$$
\begin{aligned}
& y \in (A' \cup B') \\
\Rightarrow\quad & y \in A' \text{ or } y \in B' \\
\Rightarrow\quad & y \notin A \text{ and } y \notin B \\
\Rightarrow\quad & y \notin (A \cap B) \\
\Rightarrow\quad & y \in (A \cap B)' \\
\therefore\quad & A' \cup B' \subset (A \cap B)' \\
\therefore\quad & (A \cap B)' = A' \cup B'.
\end{aligned}
$$

Theorem 1:

If A and B are any two sets, then

(i) $A - B = A \cap B'$

(ii) $B - A = B \cap A'$

(iii) $A - B = A \Leftrightarrow A \cap B = \phi$

(iv) $(A - B) \cup B = A \cup B$

(v) $(A - B) \cap B = \phi$

(vi) $A \subset B \Leftrightarrow B' \subset A'$

(vii) $(A - B) \cup (B - A) = (A \cup B) - (A \cap B)$.

Proof :

(i) Let x be an arbitrary element of A – B. Then

$$x \in A - B$$
$$\Rightarrow \quad x \in A \text{ and } y \notin B$$
$$\Rightarrow \quad x \in A \text{ and } x \in B'$$
$$\Rightarrow \quad x \in A \cap B'$$
$$\therefore \quad (A - B) \subset A \cap B' \quad \text{...(i)}$$

Again, let y be an arbitrary element of $A \cap B'$. Then

$$y \in A \cap B'$$
$$\Rightarrow \quad y \in A \text{ and } y \in B'$$
$$\Rightarrow \quad y \in A \text{ and } y \notin B$$
$$\Rightarrow \quad y \in (A - B)$$
$$\therefore \quad A \cap B' \subset A - B \quad \text{...(ii)}$$

Hence, from (i) and (ii), we have

$$(A - B) = A \cap B'.$$

(ii) Proceed as in (i).

(iii) In order to prove that $A - B \Leftrightarrow A \cap B = \phi$, we shall prove that.

(i) $A - B = A \Rightarrow A \cap B = \phi$ and

(ii) $A \cap B = \phi \Rightarrow A - B = A$.

First, let $A - B = A$. Then we have to prove that $A \cap B = \phi$. If possible, let $A \cap B \neq \phi$. Then

$$A \cap B \neq \phi \Rightarrow \text{ there exists } x \in A \cap B$$
$$\Rightarrow x \in A \text{ and } x \in B$$
$$\Rightarrow x \in A - B \text{ and } x \in B \quad [\because A - B = A]$$
$$\Rightarrow (x \in A \text{ and } x \notin B) \text{ and } x \in B$$
$$\Rightarrow x \in A \text{ and } (x \notin B \text{ and } x \in B)$$

But $x \notin B$ and $x \in B$ both can never be possible simultaneously. Thus, we arrive at a contradiction. So, our supposition is wrong.

$$\therefore \quad A \cap B = \phi.$$

Hence, $\quad A - B = A$

$$\Rightarrow \quad A \cap B = \phi. \quad \text{...(i)}$$

Conversely, let $A \cap B = \phi$. Then we have to prove that $A - B = A$. For this we shall show that $A - B \subset A$ and $A \subset A - B$.

Let x be an arbitrary element of A – B. Then

$$x \in A - B$$
$$\Rightarrow \quad x \in A \text{ and } x \notin B$$
$$\Rightarrow \quad x \in A$$
$$\therefore \quad A - B \subset A$$

Again, let y be an arbitrary element of A. Then

$$y \in A$$
$$\Rightarrow \quad y \in A \text{ and } y \notin B \qquad [\because A \cap B = \phi]$$
$$\Rightarrow \quad y \in A - B \qquad [\text{By definition of } A - B]$$
$$\therefore \quad A \subset A - B.$$
$$\text{So,} \quad A - B = A.$$
$$\therefore \quad A \cap B = \phi \Rightarrow A - B = A \qquad ...(ii)$$

Hence, from (i) and (ii), we have

$$A - B = A \Leftrightarrow A \cap B = \phi.$$

(iv) Let x be an arbitrary element of (A – B) ∪ B. Then

$$x \in (A - B) \cup B$$
$$\Rightarrow \quad x \in A - B \text{ or } x \in B$$
$$\Rightarrow \quad (x \in A \text{ and } x \notin B) \text{ or } x \in B$$
$$\Rightarrow \quad (x \in A \text{ or } x \in B) \text{ or } (x \notin B \text{ or } x \in B)$$
$$\Rightarrow \quad x \in A \cup B$$
$$\therefore \quad (A - B) \cup B \subset A \cup B$$

Let y be an arbitrary element of A ∪ B. Then

$$y \in A \cup B$$
$$\Rightarrow \quad y \in A \text{ or } y \in B$$
$$\Rightarrow \quad (y \in A \text{ or } y \in B) \text{ and } (y \notin B \text{ or } y \in B)$$
$$\Rightarrow \quad (y \in A \text{ and } y \notin B) \text{ or } y \in B$$
$$\Rightarrow \quad y \in (A - B) \cup B$$
$$\therefore \quad A \cup B \subset (A - B) \cup B$$
$$\text{Hence,} \quad (A - B) \cup B = A \cup B.$$

(v) If possible, let $(A - B) \cap B \neq \phi$. Then there exists atleast one element x, say in $(A - B) \cap B$.

$$\text{Now,} \quad x \in (A - B) \cap B$$
$$\Rightarrow \quad x \in (A - B) \text{ and } x \in B$$
$$\Rightarrow \quad (x \in A \text{ and } x \notin B) \text{ and } x \in B$$

$\Rightarrow \quad x \in A$ and $(x \notin B$ and $x \in B)$

Both $x \notin B$ and $x \in B$ both can never be possible simultaneously. Thus, we arrive at a contradiction. So, our supposition is wrong. Hence, $(A - B) \cap B = \phi$.

(vi) First, let $A \subset B$. Then we have to prove that $B' \subset A'$. Let x be an arbitrary element of B'. Then

$$x \in B'$$

$$\Rightarrow \quad x \notin B$$

$$\Rightarrow \quad x \notin A \qquad [\because A \subset B]$$

$$\Rightarrow \quad x \in A'$$

$$\therefore \quad B' \subset A' \qquad ...(i)$$

Thus, $\quad A \subset B \Rightarrow B' \subset A'$

Conversely, let $B' \subset A'$. Then we have to prove that $A \subset B$. Let y be an arbitrary element of A. Then

$$y \in A \Rightarrow y \notin A'$$

$$\Rightarrow \quad y \notin B' \qquad [\because B' \subset A']$$

$$\Rightarrow \quad y \in B$$

$$\therefore \quad A \subset B$$

Thus, $\quad B' \subset A \Rightarrow A \subset B \qquad ...(ii)$

From (i) and (ii), we have

$$A \subset B \Leftrightarrow B' \subset A'.$$

(vii) Let x be an arbitrary element of $(A - B) \cup (B - A)$. Then

$$x \in (A - B) \cup (B - A) \Rightarrow x \in (A - B) \text{ or } x \in (B - A)$$

$$\Rightarrow \quad (x \in A \text{ and } x \notin B) \text{ or } (x \in B \text{ and } x \notin A)$$

$$\Rightarrow \quad (x \in A \text{ or } x \in B) \text{ and } (x \notin B \text{ and } x \notin A)$$

$$\Rightarrow \quad x \in (A \cup B) \text{ and } x \notin (A \cap B)$$

$$\Rightarrow \quad x \in (A \cup B) - (A \cap B)$$

$$\therefore \quad (A - B) \cup (B - A) \subset (A \cup B) - (A \cup B) \qquad ...(i)$$

Again, let y be an arbitrary element of $(A \cup B) - (A \cap B)$.

Then $\quad y \in (A \cup B) - (A \cap B)$

$$\Rightarrow \quad y \in A \cup B \text{ and } y \notin A \cap B$$

$$\Rightarrow \quad (y \in A \text{ or } y \in B) \text{ and } (y \notin A \text{ and } y \notin B)$$

$$\Rightarrow \quad (y \in A \text{ and } y \notin B) \text{ or } (y \in B \text{ and } y \notin A)$$

$$\Rightarrow \quad y \in (A - B) \text{ or } y \in (B - A)$$

$\Rightarrow \quad y \in (A - B) \cup (B - A)$

$\therefore \quad (A \cup B) - (A \cap B) \subset (A - B) \cup (B - A) \quad ...(ii)$

Hence, from (i) and (ii), we have

$$(A - B) \cup (B - A) = (A \cup B) (A \cup B).$$

Theorem 2:

If A, B and C are any three sets, then

(i) $A - (B \cap C) = (A - B) \cup (A - C)$

(ii) $A - (B \cup C) = (A - B) \cap (A - C)$

(iii) $A \cap (B - C) = (A \cap B) - (A \cap C)$

(iv) $A \cap (B \Delta C) = (A \cap B) \Delta (A \cap C)$

Proof:

(i) Let x be any element of $A - (B \cap C)$.

Then $\quad x \in A - (B \cap C) \Rightarrow x \in A$ and $x \notin (B \cap C)$

$\Rightarrow \quad x \in A$ and $(x \notin B$ or $x \notin C)$

$\Rightarrow \quad (x \in A$ and $x \notin B)$ or $(x \in A$ and $x \notin C)$

$\Rightarrow \quad x \in (A - B)$ or $x \in (A - C)$

$\Rightarrow \quad x \in (A - B) \cup (A - C)$

$\therefore \quad A - (B \cap C) \subset (A - B) \cup (A - C)$

Similarly, $\quad (A - B) \cup (A - C) \subset A - (B \cap C)$

Hence, $\quad A - (B \cap C) = (A - B) \cup (A - C)$

(ii) Let x be arbitrary element of $A - (B \cup C)$. Then

$x \in A - (B \cup C) \Rightarrow x \in A$ and $x \notin (B \cup C)$

$\Rightarrow \quad x \in A$ and $(x \notin B$ and $x \notin C)$

$\Rightarrow \quad (x \in A$ and $x \notin B)$ and $(x \in A$ and $x \notin C)$

$\Rightarrow \quad x \in (A - B)$ and $x \notin (A - C)$

$\Rightarrow \quad x \in (A - B) \cap (A - C)$

$\therefore \quad A - (B \cup C) \subset (A - B) \cap (A - C)$

Similarly, $\quad (A - B) \cap (A - C) \subset A - (B \cup C)$

Hence, $\quad A - (B \cup C) = (A - B) \cap (A - C)$

(iii) Let x be any arbitrary element of $A \cap (B - C)$. Then

$x \in A \cap (B - C)$

$\Rightarrow \quad x \in A$ and $x \in (B - C)$

$\Rightarrow \quad x \in A \text{ and } (x \in B \text{ and } x \notin C)$

$\Rightarrow \quad (x \in A \text{ and } x \in B) \text{ and } (x \in A \text{ and } x \notin C)$

$\Rightarrow \quad (x \in A \cap B) \text{ and } x \notin (A \cap C)$

$\Rightarrow \quad x \in (A \cap B) - (A \cap C)$

$\therefore \quad A \cap (B - C) \subset (A \cap B) - (A \cap C)$

Similarly, $\quad (A \cap B) - (A \cap C) \subset A \cap (B - C)$

$\therefore \quad A \cap (B - C) = (A \cap B) - (A \cap C)$

(iv) $A \cap (B \,\Delta\, C) = A \cap [(B - C) \cup (C - B)]$

$= [A \cap (B - C)] \cup [A \cap (C - B)]$ (By distributive law)

$= [(A \cap B)-(A \cap C)]\cup[(A \cap C)-(A \cap B)]$ (From (iii))

$= (A \cap B) \,\Delta\, (A \cap C)$.

PRINCIPLE OF MATHEMATICAL INDUCTION

Let P(n), $n \in N$ be a statement such that (i) P(1) is true, and (ii) truth of P(r) implied the truth of P(r + 1). Then, by the principle of mathematical induction, the statement P(n) is true for all $n \in N$. Obviously, the principle of mathematical induction involves the following steps :

(i) First, we prove that the result is true for n = 1,

(ii) Then, we assume that the result is true for n = r.

(iii) Finally, we prove that the result is true for n = r + 1.

Then, we conclude by the principle of mathematical induction that the statement is true for all $n \in N$.

SOME OTHER RESULTS ON OPERATIONS ON SETS

Theorem 1:

For any sets A and B, prove that:

(i) $A - B = A \cap B'$

(ii) $(A - B) \cup B = A \cup B$.

(iii) $(A - B) \cap B = \phi$.

Proof:

(i) Let x be an arbitrary element of (A – B).

Then, $\quad x \in (A - B)$

$\Rightarrow \quad x \in A \text{ and } x \notin B$

$\Rightarrow \quad x \in A \text{ and } x \in B'$

$\Rightarrow \quad x \in (A \cap B')$.

$\therefore \quad (A - B) \subseteq (A \cap B')$

Similarly, $(A \cap B') \subseteq (A - B)$

Hence $\quad (A - B) = (A \cap B')$.

(ii) Let x be an arbitrary element of $(A - B) \cup B$

Then $x \in (A - B) \cup B$

$\Rightarrow \; x \in (A - B) \;\Rightarrow\; x \in B$

$\Rightarrow \; (x \in A \;\&\; x \notin B) \;\Rightarrow (x \in B)$

$\Rightarrow \; (x \in A \;\Rightarrow\; x \in B)$ and $(x \notin B \Rightarrow x \in B)$ [$\because$ 'or' distributes '&']

$\Rightarrow \; x \in (A \cup B)$

$\therefore \quad (A - B) \cup B \subseteq (A \cup B)$ ...(i)

Again, let y be an arbitrarily element of $(A \cup B)$. Then

$y \in (A \cup B)$

$\Rightarrow \; (y \in A \;\Rightarrow\; y \in B)$

$\Rightarrow \; (y \in A \;\Rightarrow\; y \in B)$ and $(y \notin B \;\Rightarrow\; y \in B)$

$\Rightarrow \; (y \in A \;\&\; y \notin B) \Rightarrow \; (y \in B)$ ['or' distributes '&']

$\Rightarrow \; y \in (A - B) \Rightarrow \; (y \in B)$

$\Rightarrow \; y \in (A - B) \cup B$ ...(ii)

Hence from (i) and (ii), we have $(A - B) \cup B = (A \cup B)$.

(iii) If possible, let $(A - B) \cap B \neq \phi$

and let $x \in (A - B) \cap B$.

Then, $x \in (A - B) \cap B$

$\Rightarrow \; x \in (A - B)$ and $x \in B$

$\Rightarrow (x \in A \;\&\; x \notin B)$ and $(x \in B)$

$\Rightarrow \; x \in A$ and $(x \notin A \;\&\; x \in B)$

But, $x \notin B$ and $x \in B$ both can never hold simultaneously.

Thus, we arrive at a contradiction.

Since the contradiction arises by assuming that $(A - B) \cap B \neq \phi$, hence $(A - B) \cap B \neq \phi$.

Theorem 2:

Using various laws on operations on sets, prove the following:

(i) $A \cap (B \,\Delta\, C) = (A \cap B) \,\Delta\, (A \cap C)$

(ii) $A - (A - B) = A \cap B$

(iii) $(A \cup B) - (A \cap B) = (A - B) \cap (B - A)$.

Proof:

(i) We have $A \cap (B \Delta C)$

$= A \cap [(B - C) \cup (C - B)]$

$= [A \cap (B - C)] \cup [A \cap (C - B)]$ (Distributive law)

$= [(A \cap B) - (A \cap C)] \cap (A \cap C) - (A \cap B)]$

$= (A \cap B) \Delta (A \cap C).$

(ii) We have

$A - (A - B) = A - (A \cap B')$ $[\because A - B = A \cap B']$

$= A \cap (A \cap B')'$

$= A \cap (A' \cup B')$ [by De-Morgan's law]

$= (A \cap A') \cup (A \cap B)$

$= \phi \cup (A \cap B) = A \cap B.$

(iii) $(A \cup B) - (A \cap B)$

$= (A \cup B) \cap (A \cap B)'$

$= (A \cap B) (A' \cup B')$ [De-Morgan's law]

$= [A \cap (A' \cup B')' \cup [B \cap (A' \cup B')]$ (Distributive law)

$= [(A \cap A)\cup(A \cap B')]\cup[(B \cap A')\cup(B \cap B')]$ (Distributive law)

$= [\phi \cup (A \cap B')] \cup [(B \cup A') \cup \phi]$

$= (A \cap B') \cup (B \cap A')$

$= (A - B) \cup (B - A).$

Example: *Let $A = N$ and for each $\lambda \in A$.*

Let $A_\lambda = \{1, 1/2, 1/3\}$

Let $U = \{x \in R, 0 \leq x \leq 1\}$

Find $U\{A_\lambda: \lambda \in A$.

Solution: Here $\lambda = 1, 2, 3, 4,\ldots,$ so that we have

$A_1 = \{1\}, A_2 = \{1, 1/2\}. A_3 = \{1, 1/2, 1/3\}$ etc.

it is clear that

$U\{A_\lambda: \lambda \in A\} = \{1, 1/2, 1/3,\ldots 1/\lambda, 1/\lambda+1\}$

Theorem 3:

For any sets A and B:

(i) $A \subseteq B \Leftrightarrow B' \subseteq A'.$

(ii) $A - B = A \Leftrightarrow A \cap B = \phi.$

Proof:

(i) We shall prove it in two parts, namely

$(A \subseteq B \Rightarrow B' \subseteq A')$ and $(B' \subseteq A' \Rightarrow A \subseteq B)$.

First part

Let $A \subseteq B$ be given. Then, in order to prove that $B' \subseteq A'$, let x b an arbitrary element of B'. Then

$x \in B'$

$\Rightarrow x \notin B$

$\Rightarrow x \notin A$ [$\because A \subseteq B$]

$\Rightarrow x \in A'$

$\therefore B' \subseteq A'$

Thus, $A \subseteq B \Rightarrow B' \subseteq A'$ ...(I)

Second part

Let $B' \subseteq A'$ be given. Then, in order to prove that $A \subseteq B$, let x be an arbitrary element of A. Then

$x \in A$

$\Rightarrow x \notin A'$

$\Rightarrow x \notin B'$ [$\because B' \subseteq A$]

$\Rightarrow x \in B$

$\therefore A \subseteq B$

Thus, $B \subseteq A' \Rightarrow A \subseteq B$ ...(II)

Hence, form (I) and (II), we have $A \subseteq B \Leftrightarrow B' \subseteq A'$.

(ii) We prove it in two parts, namely

First part

Let $(A - B) = A$ be given and we have to prove that, $A \cap B = \phi$. If possible, let $A \cap B = \phi$. and let $x \in A \cap B$.

Then $x \in A \cap B$

$\Rightarrow x \in A$ and $x \in B$

$\Rightarrow x \in (A - B)$ and $x \in B$ [$\because A = A - B$]

$\Rightarrow (x \in B$ and $x \notin B)$ and $(x \in B)$

$\Rightarrow x \in A$ and $(x \notin B$ & $x \in B)$

But, $x \notin B$ and $x \in B$ both can never hold simultaneously. Thus, a contradiction.

Hence $A \cap B = \phi$

Thus, $(A–B) = A \Rightarrow A \cap B = \phi$

Second part

Let A È B = f be given and in order to prove that, $A - B = A$, let x be an arbitrary element of $A - B$. Then

$x \in A - B$

$\Rightarrow x \in A$ and $x \notin B$

$\Rightarrow x \in A$ (surely)

$\therefore A - B \subseteq A$ ...(I)

Again, let y be an arbitrary element of A. Then

$y \in A$

$\Rightarrow y \in A$ and $y \notin B$ [$\because A \cap B = \phi$]

$\Rightarrow y \in A - B$

$\therefore A \subseteq A - B$ ...(II)

Thus, from (I) and (II) we have $A - B = A$

Consequently,

$A \cap B = \phi \Rightarrow A - B = A$

Hence $(A - B) = A \Leftrightarrow A \cap B = \phi$.

PRODUCTS OF SETS

We shall often have occasion to weld together the sets of a given class into a single new set called their *product* (or their *Cartesian product*). The ancestor of this concept is the coordinate plane of analytic geometry, that is, a plane equipped with the usual rectangular coordinate system. We give a brief description of this fundamental idea with a view to paving the way for our discussion of products of sets in general.

First, a few preliminary comments about the *real line*. We have already used this term several times without any explanation, and of course what we mean by it is an ordinary geometric straight line whose points have been identified with—or coordinatized by—the set R of all real numbers. We use the letter R to denote the real line as well as the set of all real numbers, and we often speak of real numbers as if they were points on the real line, and of points on the real line as if they were real numbers. Let no one be deceived into thinking that the real line is a simple thing, for its structure is exceedingly intricate. Our present view of it, however, is as naive and uncomplicated. Generally speaking, we assume that the reader is familiar with the simpler properties of the real line—those relating to inequalities and the basic algebraic operations of addition, subtraction, multiplication, and division. One of the most significant facts about the real number system is perhaps less well known. This is the so-called *least upper bound property*, which asserts that every non-empty set of real numbers which has an upper bound has a least upper bound. It is an easy consequence of this that every non-empty set of real numbers which has a lower bound has a greatest lower bound. All these matters can be developed rigorously on the basis of a small number of axioms, and detailed treatments can often be found in books on elementary abstract algebra.

To construct the coordinate plane, we now proceed as follows. We take two identical replicas of the real line, which we call the x axis and the y axis, and paste them on a plane at right angles to one another in such a way that they cross at the zero point on each. Now, let P be a point in the plane. We project P perpendicularly onto points P_x and P_y on the axes. If x and y are the coordinates of P_z and P_y on their respective axes, this process leads us from the point P to the uniquely determined ordered pair (x, y) of real numbers, where x and y are called the *x coordinate* and *y coordinate* of P. We can reverse the process, and, starting with the ordered pair of real numbers, we can recapture the point. This is the manner in which we establish the familiar one-to-one correspondence between points P in the plane and ordered pairs (x, y) of real numbers. In fact, we think of a point in the plane (which is a geometric object) and its corresponding ordered pair of real numbers (which is an algebraic object) as being—to all intents and purposes—

identical with one another. The essence of analytic geometry lies in the possibility of exploiting this identification by using algebraic tools in geometric arguments and giving geometric interpre-tations to algebraic calculations. The conventional attitude toward the coordinate plane in analytic geometry is that the geometry is the focus of interest and the algebra of ordered pairs is only a convenient tool. Here we reverse this point of view. For us, the *coordinate plane* is defined to be the set of all ordered pairs (x, y) of real numbers. We can satisfy our desire for visual images by using and by calling such an ordered pair a point, but this geometric language is more a convenience than a necessity.

Our notation for the coordinate plane is R × R, or R^2. This symbolism reflects the idea that the coordinate plane is the result of "multiplying together" two replicas of the real line R.

It is perhaps necessary to comment on one possible source of mis-understanding. When we speak of R^2 as a plane, we do so only to establish an intuitive bond with the reader's previous experience in analytic geometry. Our present attitude is that R^2 is a pure set and has no structure whatever, because no structure has yet been assigned to it. We remarked earlier (with deliberate vagueness) that a space is a set to which has been added some kind of algebraic or geometric structure. We shall convert the set R^2 into the space of analytic geometry by defining the distance between any two points (x_1, y_1) and (x_2, y_2) to be

$$\sqrt{(x_1 - x_2)^2 + (y_1 - y_2)^2}\ .$$

This notion of distance endows the set R^2 with a certain "spatial" character, which we shall recognize by calling the resulting space the *Euclidean plane* instead of the coordinate plane.

We assume that the reader is fully acquainted with the way in which the set C of all complex numbers can be identified (as a set) with the coordinate plane R^2. If 2 is a complex number, and if 2 has the standard form x + y where x and y are real numbers, then we identify z with the ordered pair (x, y), and thus with an element of R^2. The complex numbers, however, are much more than merely a set. They constitute a number system, with operations of addition, multiplication, conjugation, etc. When the coordinate plane R^2 is thought of as consisting of complex numbers and is enriched by the algebraic structure it acquires in this way, it is called the *complex plane*. The letter C is used to denote either the set of all complex numbers or the complex plane.

Suppose now that X_1 and X_2 are any two non-empty sets. By analogy with our above discussion, their *product* $X_1 \times X_2$ is defined to be the set of all ordered pairs $(x_1, x_2\}$, where x_1 is in X_1 and x_2 is in X_2. In spite, of the arbitrary nature of X_1 and X_2, their product can be represented by a picture (Fig. 1.8) which is loosely similar to the usual picture of the coordinate plane. The term *product* is applied to this set, and it is thought of as the result of "multiplying together" X_1 and X_2, for the following reason: if X_1 and X_2 are finite sets with m and n elements, then (clearly) $X_1 \times X_2$ has mn elements. If f: $X_1 \to X_2$ is a mapping with domain X_1 and range in X_2, its *graph* is that subset of $X_1 \times X_2$ which consists of all ordered pairs of the form $(x_1, f(x_1))$. We observe that this is an appropriate generalization of the concept of the graph of a function as it occurs in elementary mathematics.

This definition of the product of two sets extends easily to the case of n sets for any positive integer n. If $X_1, X_2, \ldots, X_n$ are non-empty sets, then their product $X_1 \times X_2 \times \ldots \times X\ X_n$ is the set of all ordered n-tuples $(x_1, x_2, \ldots, x_n)$, where x_i is in X_i for each subscript i. If the X_i's are all replicas of a single set X, that is, if

$$X_1 = X_2 = \ldots. = X_n = X,$$

then their product is usually denoted by the symbol X^n.

These ideas specialize directly to yield the important sets R^n and C^n. R^1 is just R, the real line, and R^2 is the coordinate plane, R^3 — the set of all ordered triples of real numbers—is the set which underlies solid analytic geometry, and we assume that the reader is familiar with the manner in which this set arises, through the introduction of a rectangular coordinate system into ordinary three-dimensional space. We can draw pictures here just as in the case of the coordinate plane, and we can use geometric language as much as we please, but it must be understood that the mathematics of this set is the mathematics of ordered triples of real numbers and that the pictures are merely an aid to the intuition. Once we fully grasp this point of view, there is no difficulty whatever in advancing at once to the study of the set R^n of all ordered n-tuples $(x_1, x_2, \ldots, x_n)$ of real numbers for any positive integer n. It is quite true that when n is greater than 3 it is no longer possible to draw the same kinds of intuitively rich pictures, but at worst this is merely an inconvenience. We can (and do) continue to use suggestive geometric language, so all is not lost. The set C^n is defined similarly: it is the set of all ordered n-tuples $(z_1, z_2, \ldots, z_n)$ of complex numbers. Each of the sets R^n and C^n plays a prominent part in our later work.

We emphasized above that for the present the coordinate plane is to be considered as merely a set, and not a space. Similar remarks apply to R^n and C^n. In due course] we shall impart form and content to each of these sets by suitable definitions. We shall convert them into the *Eucliaean and Unitary n-spaces* which underlie and motivate so many developments in modern pure mathematics. But as of now—and this is the point we insist on—neither one of these sets has any structure *at all*.

As the reader doubtless suspects, it is not enough that we consider only products of finite classes of sets. The needs of topology compel us to extend these ideas to arbitrary classes of sets.

We defined the product $X_1 \times X_2 \times .... \times X_n$ to be the set of all ordered n-tuples $(x_1, x_2, ..., x_n)$ such that x_1 is in X_i for each subscript i. To see how to extend this definition, we reformulate it as follows. We have an index set 7, consisting of the integers from 1 to n, and corresponding to each index (or subscript) i we have a non-empty set X_i. The n-tuple $(x_1, x_2, ... , x_n)$ is simply a function (call it x) defined on the index set I, with the restriction that its value $x(i) = x_i$ is an element of the set X_i for each i in I. Our point of view here is that the function x is completely determined by, and is essentially equivalent to, the array $(x_i, x_2, ... , X_n)$ of its values.

The way is now open for the definition of products in their full generality. Let $\{X_1\}$ be a non-empty class of non-empty sets, indexed by the elements i of an index set I. The sets X_i need not be different from one another; indeed, it may happen that they are all identical replicas of a single set, distinguished only by different indices. The *product* of the sets X_i, written $P_{ieI} X_i$, is defined to be the set of all functions x defined on I such that x(i) is an element of the set X_i for each index i. We call X, the *ith coordinate set*. When there can be no misunderstanding about the index set, the symbol P_{ieI} X, is often abbreviated to P_iX_i. The definition we have just given requires that each coordinate set be non-empty before the product can be formed. It will be useful if we extend this definition slightly by agreeing that if any of the X_i's are empty, then P,X, is also empty.

This approach to the idea of the product of a class of sets, by means of functions defined on the index set, is useful mainly in giving the definition. In practice, it is much more convenient to use the subscript notation x_i instead of the function notation (x_i). We then interpret the product P_iX_i as made up of elements x, each of which is specified by the exhibited array $\{x_i\}$ of its values in the respective coordinate sets X_i. We call x_i the *ith coordinate* of the element $x = \{x_1\}$.

The mapping p_i of the product P_iX_i onto its tth coordinate set X_i which is defined by $p_i(x) = x_i$—that is, the mapping whose value at an arbitrary element of the product is the ith coordinate of that element—is called the projection onto the tth coordinate set. The projection p_i selects the zth coordinate of each element in its domain. There is clearly one projection for each element of the index set I, and the set of all projections plays an important role in the general theory of topological spaces.

If A and B are any two sets, then set of all distinct ordered pairs whose first coordinate is an element of A and whose second coordinate is an element of B is called the *Cartesian product.* A and B and is denoted by $A \times B$.

Symbolically, $A \times B = \{a, b) : a \in A \text{ and } b \in B\}$

For example, if $A = \{a, b, c\}$ and

$B = \{x, y, z)$ then

$$A \times B = \{(a, x), (a, y), (a, z), (b, x) (b, y), (b, z), (c, x), (c, y), (c, z)\}$$

and $$B \times A = \{(x, a), (x, b), (x, c), (y, a) (y, b), (y, c), (z, a), (z, b), (z, c)\}.$$

Notes:

(1) $A \times B \neq B \times A$, *i.e.*, cartesian product is not commutative.

(2) If the set A and B have m and n elements respectively, then the set $A \times B$ has *mn* elements.

(3) *If either A or B is a null set then the set $A \times B$ is also a null set.*

(4) *If either A or B is an infinite set and other is a non-empty set, then $(A \times B)$ is also an infinite set.*

(5) $A \times B = B \times A \Rightarrow A = B$.

RELATIONS

The word relation is not unfamiliar even to a child. He is aware of his various relations with family members. As he grows up he comes across the other types of relations also, *e.g.* relation between money and its purchasing power, relation of education with school etc. The description of relation in above language is very vague. Here we shall try to define and describe relation in the precise language of mathematics.

First we shall restrict ourselves only to the relations between only two objects. The relations which associate only two objects are known as *'Binary relation'*. Below we give a mathematical definition of a relation.

Definition : *Let A and B be two sets. A relation from A to B is a subset of $A \times B$. Symbolically, R is a relation from A to B iff $R \subset A \times B$.*

If (x, y) be a member of a relation set R, we express it by writing x R y and say that 'x is in the relation R to y'. Thus

$$(x, y) \in R \Leftrightarrow x\ R\ y.$$

For example, if $A = \{2, 3, 5, 6\}$ and R means 'divides' then

2R2, 2R6, 3R3, 3R6, 5R5, 6R6 and as such relation set $R = \{(2, 2), (2, 6), (3, 3) (3, 6), (5, 5), (6, 6)\}$.

Furthermore if I be the set of all integers, the statement "x is less than y where x, y ∈ I" determines a relation in I. If we denote this relation by R, then we may describe the set R in the set builder notation as given below.

$$R = \{(x, y) : x, y \in I, x < y\}$$

Binary Relation : A relation R between pairs of element of a set A is called a *binary relation.* In the discussions here after the word relation will mean binary relation.

Let A be any set and R be the set of all those pairs (x, y) of A × A for which x = y. Then the relation R is called the equality relation in the set A. If is also called the *diagonal relation* of A and is denoted by Δ. Thus

$$\Delta = \{(x, y) : x \in A\}.$$

DOMAIN AND RANGE OF A RELATION

Let R be a relation from A to B. Then the set of all first co-ordinates of the members of the relation set R is called the *domain* of R and the set of all second members of R is called the *range* of R. Thus

$$\text{Domain } R = \{x : (x, y) \in R\}$$

$$\text{Range } R = \{y : (x, y) \in R\}$$

Fore example if A = {1, 2, 3, 4}, B = {3, 4, 5}, then

$$A \times B = \{(1, 3), (1, 4), (1, 5). (2, 3), (2, 4)$$
$$(2, 5), (3, 3), (3, 4). (3, 5), (4, 3)$$
$$(4, 4), (4, 5)\}.$$

Now, if R stands for 'less than', then relation set

$$R = \{(1, 3), (1, 4), (1, 5). (2, 3), (2, 4)$$
$$(2, 5), (3, 4). (3, 5), (4, 5).$$

Domain R = set of first co-ordinates of ordered pairs in the relation set = {1, 2, 3, 4}

Range R = set of second co-ordinates of ordered pairs in the relation set = {3, 4, 5}

COMPOSITION OF RELATIONS

Let A, B, and C be sets, and let R be a relation from A to B and let S be a relation from B to C. Obviously, R is a subset of A × B and S is a subset of B × C. Then R and S give rise to a relation from A to C denoted by RoS and defined by a (RoS) c if for some b ∈ B we have a R b and b S c.

Symbolically :

RoS = {(a, c) : ∀ b ∈ B for which (a, b) ∈ R and (b, c) ∈ S}

Relation RoS is called the *composition of* R and S. It is sometimes denoted as RS.

Let R is a relation on a set A. Then RoR, the composition of R with itself is always defined, and is sometimes denoted by R^2. Similarly, $R^3 = R^2oR$ = RoRoR and so on. Thus R^n is defined for all positive n.

Theorem:

Let A, B, C and D be sets. Suppose R is a relation from A to B, S is a relation from B to C and T is a relation from C to D. Then show that

(RoS) oT = Ro (SoT)

Proof :

Here, we need to show than each ordered pair in (RoS) oT belongs to Ro (SoT) and vice versa.

Let (a, d) belongs to (RoS) oT.

Then, there exists a c in C such than (a, c) ∈ RoS and (c, d) ∈ T.

Since (a, c) ∈ RoS, there exists a b in B such that (a, b) ∈ R and (b, c) ∈ S.

Since (b, c) ∈ S and (c, d) ∈ T,

we have (b, d) ⊂ SoT.

Again, since (a, b) ∈ R and (b, d) ∈ SoT,

we have (a, d) ∈ Ro (SoT).

Therefore, (RoS) oT ⊂ Ro (SoT) . . . (1)

Similarly, Ro (SoT) ⊂ (RoS) oT . . . (2)

From (1) and (2), we get

Ro (SoT) = (RoS) oT.

TOTAL NUMBER OF RELATIONS

Let A and B be two non-empty finite sets consisting of m and n elements respectively. Then A × B consists of mn ordered pairs, So, total number of subsets of A × B is 2^{mn}. Since each subset of A × B defines a relation from A to B, so total number of relations from A to B is 2^{mn}. Among these 2^{mn} relations the void relation ϕ and the universal relation A × B are trivial relations from A to B.

Relation on a Set : Let A be a non-void set. Then, a relation of A × A, is called relation on set A.

Inverse Relation : Let A, B be two sets and let R be a relation from a set A to a set B. Then the inverse of R, denoted by R^{-1}, is a relation from B to A and is defined by

$$R^{-1} = \{(b, a) : (a, b) \in R\}.$$

Clearly, $(a, b) \in R \Leftrightarrow (b, a) \in R^{-1}$

Also Dom $(R) = \text{Range } (R^{-1})$ and

Range $(R) = \text{Dom } (R^{-1})$.

TYPES OF RELATIONS

Void Relation : Let A be as set. Then $\phi \subset A \times A$ *and so it is a relation on A. This relation is called the void or empty relation on A.*

Universal Relation : Let A be a set. Then $A \times A \subset A \times A$ and so it is a relation on A. This relation is called the *universal relation* on A.

Note : Void and the universal relations on a set A are respectively the smallest and the largest relations on A.

Identity Relation : Let A be a set. Then the relation $I_A = \{(a, a) : a \in A\}$ on A is called the *identity relation* on A.

Obviously, the relation I_A on A is called the identity relation if every element of A is related to itself only.

Reflexive Relation : *A relation R on a set A is said to be reflexive if every element of A is related to itself.* Thus, R is reflexive $\Leftrightarrow (a, a) \in R$ for all $a \in A$. A relation R on a set A is not reflexive if there exists an element $a \in A$ such that $(a, a) \notin R$.

Example :

(i) The universal relation on a non-void set A is reflexive.

(ii) The relation R on N defined by $(x, y) \in R \Leftrightarrow x \geq y$ is a reflexive relation on N, because every natural number is greater than or equal to itself.

Symmetric Relation : *A relation R on a set A is said to be a symmetric relation iff (a, b)* $\in$ *R* $\Rightarrow$ *(b, a)* $\in$ R for all a, b $\in$ A.

i.e. $a R b \Rightarrow b R a$ for all $a, b \Rightarrow A$.

Example :

(i) The identity and the universal relation on a non-void set are symmetric relations.

(ii) Let L be the set of all lines in a plane and let R be a relation defined on L by the rule $(x, y) \in R \Leftrightarrow x$ is perpendicular to y. Then R is a symmetric relation on L, because $L_1 \perp L_2 \Rightarrow L_2 \perp L_1$, *i.e.*, $(L_1, L_2) \in R \Rightarrow (L_2, L_1) \in R$.

Note : A relation R on set A is not a symmetric relation if there are at least two elements $a, b \in A$ such that $(a, b) \in R$ but $(b, a) \notin R$.

Transitive Relation : *Let A be any set. A relation R on A is said to be a transitive relation iff* $(a, b) \in R$ *and* $(b, c) \in R \Rightarrow (a, c) \in R$ *for all* $a, b, c \in R$,

i.e. $a\, R\, b$ *and* $b\, R\, c \Rightarrow a\, R\, c$ *for all* $a, b, c, \in A$.

Example :

(i) The relation R on the set N of all natural numbers defined by $(x, y) \in R \Leftrightarrow x$ divides y, for all $x, y \in N$ is transitive.

(ii) On the set N of natural numbers, the relation R defined by $x\, R\, y \Rightarrow$ x is less than y is transitive, because for any $x, y, z \in N$

$x < y$ and $y < z \Rightarrow x < z \Rightarrow$

$x\, R\, y$ and $y\, R\, z \Rightarrow x\, R\, z$.

Antisymmetric Relation : *Let A be any set. A relation R on set A is said to be an antisymmetric relation iff*

$(a, b) \in R$ *and* $(b, a) \in R \Rightarrow a = b$ for all $a, b \in A$.

Examples :

(i) The identity relation on a set A is an antisymmetric relation.

(ii) Let R be a relation on the set N of natural numbers defined by

$x\, R\, y \Leftrightarrow$ 'x divides y' for all $x, y \in N$.

This relation is an antisymmetric relation on N. Since for any two numbers $a, b \in N$

$$a \mid b \text{ and } b \mid a \Rightarrow a = b, \text{ i.e., } a\, R\, b \text{ and } b\, R\, a \Rightarrow a = b.$$

Note : If $(a, b) \in R$ and $(b, a) \notin R$, then also R is an antisymmetric relation.

Equivalence Relation : *A relation R on a set A is said to be an equivalence relation on A iff*

(i) it is reflexive, i.e. $(a, a) \in R$ *for all* $a \in A$.

(i) it is symmetric, i.e. $(a, b) \in R \Rightarrow (b, a) \in R$ *for all* $a, b \in A$.

(i) it is transitive, i.e. $(a, b) \in R$ *and* $(b, c) \in R \Rightarrow (a, c) \in R$ *for all*

$a, b, c, \in A.$

PARTITIONS AND EQUIVALENCE RELATIONS

In the first part of this section we consider a non-empty set X, and we study decompositions of X into non-empty subsets which fill it out and have no elements in common with one another. We give special attention to the tools (equivalence relations) which are normally used to generate such decompositions.

A *partition* of X is a disjoint class $\{X_i\}$ of non-empty subsets of X whose union is the full set X itself. The X_i's are called the *partition sets*. Expressed somewhat differently, a partition of X is the result of splitting it, or subdividing it, into non-empty subsets in such a way that each element of X belongs to one and only one of the given subsets.

If X is the set {1, 2, 3, 4, 5}, then {1, 3, 5}, {2, 4} and {1, 2, 3}, {4, 5} are two different partitions of X. If X is the set R of all real numbers, then we can partition X into the set of all rationals and the set of all irrationals, or into the infinitely many closed-open intervals of the form [n, n + 1) where n is an integer. If X is the set of all points in the coordinate plane, then we can partition X in such a way that each partition set consists of all points with the same x coordinate (vertical lines), or so that each partition set consists of all points with the same y coordinate (horizontal lines).

Other partitions of each of these sets will readily occur to the reader. In general, there are many different ways in which any given set can be partitioned. These manufactured examples are admittedly rather uninspiring and serve only to make our ideas more concrete. Later in this section we consider some others which are more germane to our present purposes.

A *binary relation* in the set X is a mathematical symbol or verbal phrase, which we denote by R in this paragraph, such that for each ordered pair (x, y) of elements of X the statement x R y is meaningful, in the sense that it can be classified definitely as true or false. For such a binary relation, x R y symbolizes the assertion that x is related by R to y, and x R y the negation of this, namely, the assertion that x is *not* related by R to y. Many examples of binary relations can be given, some familiar and others less so, some mathematical and others not. For instance, if X is the set of all integers and R is interpreted to mean "is less than," which of course is usually denoted by the symbol <, then we clearly have $4 < 7$ and $5 \nless 2$. We have been speaking of binary relations, which are so named because they apply only to ordered pairs of elements, rather than to ordered triples, etc. In our work we drop the qualifying adjective and speak simply of a *relation* in X, since we shall have occasion to consider only relations of this kind.'

We now assume that a partition of our non-empty set X is given, and we associate with this partition a relation in X. This relation is defined in the following way: we say that x is *equivalent* to y and write this $x \sim y$ (the symbol $\sim$ is pronounced "wiggle"), if x and y belong to the same partition set. It is obvious that the relation $\sim$ has the following properties:

(1) $x \sim x$ for every x (reflexivity);

(2) $x \sim y \Rightarrow y \sim x$ (symmetry);

(3) $x \sim y$ and $y \sim z \Rightarrow x \sim z$ (transitivity).

This particular relation in X arose in a special way, in connection with a given partition of X, and its properties are immediate consequences of its definition. Any relation whatever in X which possesses these three properties is called an *equivalence relation* in X.

We have just seen that each partition of X has associated with it a natural equivalence relation in X. We now reverse the situation and show that a given equivalence relation in X determines a natural partition of X.

Let $\sim$ be an equivalence relation in X; that is, assume that it is reflexive, symmetric, and transitive in the sense described above. If x is an element of X, the subset of X defined by $[x] = \{y{:}y \sim x\}$ is called the *equivalence set* of x. The equivalence set of x is thus the set of all elements which are equivalent to x. We show that the class of all distinct equivalence sets forms a partition of X. By reflexivity, $x \varepsilon [x]$ for each element x in X, so each equivalence set is non-empty and their union is X. It remains to be shown that any two equivalence sets $[x_1]$ and $[x_2]$ are either disjoint or identical. We prove this by showing that if $[x_1]$ and $[x_2]$ are not disjoint, then they must be identical. Suppose that $[x_1]$ and $[x_2]$ are not disjoint; that is, suppose that they have a common element z. Since 2 belongs to both equivalence sets, $z \sim x_1$ and $a \sim x_2$, and by symmetry, $x_1 \sim z$. Let y be any element of $[x_1]$, so that $y \sim x_1$. Since $y \sim x_1$ and $x_i \sim z$, transitivity shows that $y \sim z$. By another application of transitivity, $y \sim z$ and $z \sim x_2$ imply that $y \sim x_2$, so that y is in $[x_2]$. Since y was chosen arbitrarily in [xi], we see by this that $[x_1] \subseteq [x_2]$. The same reasoning shows that $[x_2] \subseteq [x_1]$, and from this we conclude (see the last paragraph of Sec. 1) that $[x_1] = [x_2]$.

The above discussion demonstrates that there is no real distinction (other than a difference in language) between partitions of a set and equivalence relations in the set. If we start with a partition, we get an equivalence relation by regarding elements as equivalent if they belong to the same partition set, and if we start with an equivalence relation, we get a partition by grouping together into subsets all elements which are equivalent to one another. We have here a single mathematical idea, which we have been considering from two different points of view, and the approach we choose in any particular application depends entirely on our own convenience. In practice, it is almost invariably the case that we use equivalence relations (which are usually easy to define) to obtain partitions (which are sometimes difficult to describe fully).

We now turn to several of the more important simple examples of equivalence relations.

Let I be the set of all integers. If a and b are elements of this set, we write a = b (and say that a *equals* b) if a and 6 are the same integer. Thus, 2 + 3 = 5 means that the expressions on the left and right are simply different ways of writing the same integer. It is apparent that = used in this sense is an equivalence relation in the set I:

(1) a = a for every a;

(2) a = b ⇒ b = a;

(3) a = b and b = c ⇒ a = c.

Clearly, each equivalence set consists of precisely one integer.

Another familiar example is the relation of equality commonly used for fractions. We remind the reader that, strictly speaking, a fraction is merely a symbol of the form a/b , where a and b are integers and b is not zero. The fractions 2/3 and 4/6 are obviously not identical, but nevertheless we consider them to be equal. In general, we say that two fractions a/b and c/d are *equal*, written a/b = c/d, it ad and be are equal as integers in the usual sense (see the above paragraph). We leave it to the reader to show that this is an equivalence relation in the set of all fractions. An equivalence set of fractions is what we call a *rational number*. Everyday usage ignores the distinction between fractions and rational numbers, but it is important to recognize that from the strict point of view it is the rational numbers (and not the fractions) which form part of the real number system.

Our final example has a deeper significance, for it provides us with the basic tool for our work of the next two sections.

For the remainder of this section we consider a relation between pairs of non-empty sets, and each set mentioned (whether we say so explicitly or not) is assumed to be non-empty. If X and Y are two sets, we say that X is *numerically equivalent* to Y if there exists a one-to-one correspondence between X and Y, *i.e.*, if there exists a one-to-one mapping of X onto Y. This relation is reflexive, since the identity mapping $i_X : X \to X$ is one-to-one onto; it is symmetric, since it $f: X \to Y$ is one-to-one onto, then its inverse mapping $f^{-1}: Y \to X$ is also one-to-one onto; and it is transitive, since it $f: X \to Y$ and $g: Y \to Z$ are one-to-one onto, then $gf : X \to Z$ is also one-to-one onto.

Numerical equivalence has all the properties of an equivalence relation, and if we consider it as an equivalence relation in the class of all non-empty subsets of some universal set U, it groups together into equivalence sets all those subsets of U ä which have the *same number of elements*. After we state and prove the following very useful but rather technical theorem.

The theorem we have in mind—the *Schroeder-Bernstein theorem*—is the following: if X and Y *are two sets each of which is numerically equivalent to a subset of the other, then all of X is numerically equivalent to all of Y.* There are several proofs of this classic theorem, some of which are quite difficult. The very elegant proof we give is essentially due to Birkhoff and MacLane.

Now for the proof. We assume that $f:X \rightarrow Y$ is a one-to-one mapping of X into Y, and that $g : Y \rightarrow X$ is a one-to-one mapping of Y into X. Our task is to produce a mapping $F:X \rightarrow Y$ which is one-to-one onto. We may assume that neither f nor g is onto, since if f is, we can define F to be f, and if g is, we can define F to be g^{-1}. Since both f and g are one-to-one, it is permissible to use the mappings f^{-1} and g^{-1} as long as we clearly understand that f^{-1} is defined only on f(X) and g^{-1} only on g(Y). We obtain the mapping F by splitting both X and Y into subsets which we characterize in terms of the ancestry of their elements.

Let x be an element of X. We apply g^{-1} to it (if we can) to get the element $g^{-1}(x)$ in Y. If $g^{-1}(x)$ exists, we call it the first ancestor of x. The element x itself we call the zeroth ancestor of x. We now apply f^{-1} to $g^{-1}(x)$ if we can, and if $(f^{-1}g^{-1})$ (x) exists, we call it the second ancestor of x. We now apply g^{-1} to $(f^{-1}g^{-1})$ (x) it we can, and if $(g^{-1}f^{-1}g^{-1})(x)$ exists, we call it the third ancestor of x. As we continue this process of tracing back the ancestry of x, it becomes apparent that there are three possibilities.

(1) x has infinitely many ancestors. We denote by X_i the subset of X which consists of all elements with infinitely many ancestors.

(2) x has an even number of ancestors; this means that x has a last ancestor (that is, one which itself has no first ancestor) in X. We denote by X_e the subset of X consisting of all elements with an even number of ancestors.

(3) x has an odd number of ancestors; this means that x has a last ancestor in Y. We denote by X_o the subset of X which consists of all elements with an odd number of ancestors. The three sets X_i, X_e, X_o form a disjoint class whose union is X. We decompose Y in just the same way into three subsets Y_i, Y_e, Y_o. It is easy to see that f maps X_i onto Y_i and Xe onto Y_o, and that g^{-1} maps X_o onto Y_e; and we complete the proof by defining F in the following piecemeal manner:

$$F(x) = \begin{cases} f(x) & \text{if } x \varepsilon X_i \cup X_e, \\ g^{-1}(x) & \text{if } x \varepsilon X_o, \end{cases}$$

We attempt to illustrate these ideas in Fig. 1.10. Here we present two replicas of the situation: on the left, X and Y are represented by the vertical lines, and f and g by the lines slanting down to the right and left; and on the right, we schematically trace the ancestry of three elements in X, of which x_1 has no first ancestor, x_2 has a first and second ancestor, and X_2 has a first, second, and third ancestor.

The Schroeder-Bernstein theorem has great theoretical and practical significance. Its main value for us lies in its role as a tool by means of which we can prove numerical equivalence with a minimum of effort for many specific sets.

DEDUCTION AND INDUCTION

Deduction : The process of deducing particular results from a general result is called *deduction.* For example, we know that the sum of the first n natural numbers is $\frac{n(n+1)}{2}$. From this result, we deduce that the sum of the first 50 natural numbers is 1275.

Induction : The process of establishing a valid general result from particular results is called *induction.* The principle of mathematical indiction is used to establish the validity of a general result involving natural numbers.

Consider the following statement :

$$P(n) = \text{“}n(n + 1) \text{ is even”}.$$

When $\quad n = 2$, we have $P(2) = 2(3) = 6$ is even.

We cannot continue like this by testing for all natural numbers $n \in N$. We can prove that the result is true by using the following argument. Let n be even. Then n (n + 1) is even. Let n be odd. Then (n + 1) is even. Therefore n(n +1) is even. The result is proved. However, all the statements as given above may not be as simple to prove.

SOME THEOREMS ON EQUIVALENCE RELATIONS

Theorem 1:

The union of two equivalence relations on a set is not necessarily an equivalence relation on the set.

Proof:

Let A = {a, b, c} and let R and S be two relations on A, given by

$$R = \{(a, a), (b, b), (c, c), (a, b), (b, a)\}$$

and $$S = \{(a, a), (b, b), (c, c), (b, c), (c, b)\}$$

It can be easily seen that each one of R and S is an equivalence relation on A. But $R \cup S$ is not transitive, because $(a, b) \in R \cup S$ and $(b, c) \in R \cup S$ but $(a, c) \notin R \cup S$. Hence, $R \cup S$ is not an equivalence relation on A.

Theorem 2:

If R is an equivalence relation on a set A, then R^{-1} is also an equivalence relation on A.

Or

The inverse of an equivalence relation is an equivalence relation.

Proof:

Since R is a relation on A. So, $R \subset A \times A \Rightarrow R^{-1} \subset A \times A \Rightarrow R^{-1}$ is also a relation on A.

Now, we shall show that R^{-1} is an equivalence relation on A.

Reflexivity : Let a be an arbitrary element of A. Then.

$a \in A \Rightarrow (a, a) \in R$ $(\because$ R is reflexive)

$\Rightarrow (a, a) \in R^{-1}$

Thus, $(a, a) \in R^{-1}$ for all $a \in A$. So, R^{-1} is reflexive on A.

Symmetry : Let $(a, b) \in R^{-1}$, Then

$(a, b) \in R^{-1} \Rightarrow (b, a) \in R$ (By def. of R^{-1})

$\Rightarrow (a, b) \in R$ $(\because$ R is symmetric)

$\Rightarrow (b, a) \in R^{-1}$ (By def. of R^{-1})

Thus, $(a, b) \in R^{-1}$

$\Rightarrow (b, a) \in R^{-1}$ for all $a, b \in A$, So, R^{-1} is symmetric on A.

Transitivity : Let $(a, b) \in R^{-1}$

and $(b, c) \in R^{-1}$. Then

$(a, b) \in R^{-1}$ and $(b, c) \in R^{-1}$

$\Rightarrow (b, a) \in R$ and $(c, b) \in R$ (By def. of R^{-1})

$\Rightarrow (c, b) \in R$ and $(b, a) \in R$

$\Rightarrow (c, a) \in R$ $(\because$ R is transitive)

$\Rightarrow (a, c) \in R^{-1}$ (By def. of R^{-1})

Thus, $(a, b) \in R^{-1}$ and $(b, c) \in R^{-1}$

$\Rightarrow (a, c) \in R^{-1}$ for all a, b, c, $\in$ A. So R^{-1} is transitive on A.

Hence, R^{-1} is an equivalence relation on A.

Theorem 3:

If R and S are two equivalence relations on a set A then $R \cap S$ is also an equivalence relation on A.

Or

The intersection of two equivalence relations on a set is an equivalence relation on the set.

Proof:

It is given that R and S are relations on set A.

$\therefore \quad R \subset A \times A$ and $S \subset A \times A$

$\Rightarrow \quad R \cap S \subset A \times A$

$\Rightarrow \quad R \cap S$ is also a relation on A.

Now, we shall show that it is an equivalence relation on A.

We observe the following properties:

Reflexivity : Let a be an arbitrary element of A. Then.

$a \in A \quad \Rightarrow (a, a) \in R$ and $(a, a) \in S$ [$\because$ R and S are reflexive]

$\Rightarrow (a, a) \in R \cap S$

Thus, $(a, a) \in R \cap S$ for all $a \in A$, so, $R \cap S$ is a reflexive relation on A.

Symmetry : Let a, b, $\in$ A such that $(a, b) \in R \cap S$. Then

$(a, b) \in R \cap S \Rightarrow (a, b) \in R$ and $(a, b) \in S$

$\Rightarrow (b, a) \in R$ and $(b, a) \in S$

($\because$ R and S are symmetric)

$\Rightarrow (b, a) \in R \cap S$.

Thus, $(a, b) \in R \cap S \Rightarrow (b, a) \in R \cap S$ for all $(a, b) \in R \cap S$. So, $R \cap S$ is symmetric an A.

Transitivity : Let a, b, c $\in$ A such that $(a, b) \in R \cap S$ and $(b, c) \in R \cap S$. Then, $(a, b) \in R \cap S$ and $(b, c) \in R \cap S$.

$\Rightarrow$ {$(a, b) \in R$ and $(a, b) \in S$)} and {$(b, c) \in R$ and $(b, c) \in S$)}

$\Rightarrow$ {$(a, b) \in R$ and $(b, c) \in R$)} and {$(a, b) \in S$ and $(b, c) \in S$)}

$\Rightarrow$ $(a, c) \in R$ and $(a, c) \in S$

($\because$ R and S are transitive, so, $(a, b) \in R$ and $(b, c) \in R \Rightarrow (a, c) \in R$)

$\Rightarrow (a, c) \in R \cap S$

Thus, $(a, b) \in R \cap S$ and $(b, c) \in R \cap S$

$\Rightarrow (a, c) \in R \cap S$. So, $R \cap S$ is transitive on A.

Hence, $R \cap S$ is an equivalence relation on A.

CLOSURE PROPERTIES

Let A is a set. Consider the collection of all relations on A. Let P be a property of such relations such as being symmetric or being transitive. A relation with property P will be called a P-relation.

P will be called *R-closable* if P satisfies the following two conditions.

(i) There is a P-relation S containing R.

(ii) The intersection of P-relations is a P-relation.

P-closure of an arbitrary relation R on A, written P (R), is a P-relation such that,

$$R \subseteq P(R) \subseteq S.$$

Reflexive, Symmetric and Transitive Closures : Let $\Delta_A = \{(a, a) : a \in A\}$ is the equality relation on A. Let R be a relation on set A, then.

(i) $R \cup R^{-1}$ is symmetric closure of R

(ii) $R \cup \Delta_A$ is reflexive closure of R.

(iii) $R^* = \bigcup_{i=1}^{\infty} R^i$ is the transitive closure of R.

EQUIVALENCE CLASSES

Let A be any non-empty set and let R be an equivalence relation in A. Further let a be an arbitrary element of A for which a R b, is known as equivalence class of a. We shall denote this equivalence class by [a] or $\bar{a}$. Evidently *an equivalence class is a subset of the given set, such that any two elements on it are equivalent to each other.*

Some authors denote the equivalence class of a determined by a relation R in A by the symbol a/R. The set of all equivalence classes will be denoted by A/R and read as *"A modulo R". This* is called the quotient set of A by R.

Thus the set of all disjoint equivalence classes defined by an equivalence relation R over a set A is called the Quotient set of A relative to R and also denoted by $\overline{A}$.

Example : Let A be the set of all triangles in a plane and let R be an equivalence relation in A defined by "x is congruent to y", $x, y \in A$. When $x \in A$ we shall mean by the equivalence class [x] the set of all triangles of A congruent to the triangle x. Similarly, when $y \in A$ we shall mean by the equivalence class [y] the set of all triangles of A congruent to the triangle y.

Properties of Equivalence Classes

Let A be non-empty set and let R be an equivalence relation in A. Let x and y be arbitrary element in A. Then

(i) $x \in [x]$,

(ii) If $y \in [x]$, then $[y] = [x]$

(iii) $[x] = [y] \Leftrightarrow (x, y) \in R$ *i.e.* iff x R y.

(iv) Either $[x] = [y]$ or $[x] \cap [y] = \phi$,

i.e. two equivalence classes are either disjoint or identical.

PARTITIONS OF A SET

Let X be a non-empty set. A set $P = \{A, B, C\}$ of non-empty subsets of X will be called a partition of X if

(i) $A \cup B \cup C \ldots = X$, *i.e.* the set X is the union of the sets in P and

(ii) The intersection of every pair of distinct subsets of $X \in P$ is the null set *i.e.* if A and $B \in P$ then either $A = B$ or $A \cap B = \phi$.

Example : *Consider the set* $X = \{1, 2, \ldots .9, 10\}$ *and its subsets* $B_1 = \{1, 3\}$, $B_2 = \{7, 8, 10\}$, $B_3 = \{2, 5, 6\}$, $B_4 = \{4, 9\}$

The set $P = \{B_1, B_2, B_3, B_4\}$ is such that

(i) B_1, B_2, B_3, B_4 are all non-empty subsets of X.

(ii) $B_1 \cup B_2 \cup B_3 \cup B_4 = X$.

(iii) For any sets B_i, $B_i \cap B_j = \phi$.

Hence the set $\{B_1, B_2, B_3, B_4\}$ is a partition of X.

***Theorem 1*:**

An equivalence relation defined in a set de composes the set into disjoint equivalence classes.

Proof :

Let an equivalence relation R be defined in a set S. Let $a \in S$ and T be a subset of S consisting of all those elements which are equivalent to a, *i.e.*

$$T = \{x : x \in S \text{ and } x \text{ R } a\}$$

Then $a \in T$, for a R a (R is reflexive).

Any two elements of T are equivalent to each other, for if $x, y \in T$.

Then x R a and y R a.

Again x R a, y R a

$\Rightarrow$ x R a, a R y (R is symmetric)

$\Rightarrow$ x R y (R is transitive)

Thus T is an equivalence class.

Let T_1 be another equivalence class, *i.e.*

$$T_1 = \{x : x \in S \text{ and } x R b\},$$

where b is not equivalent to a. Then the class T and T_1 must be disjoint. For if they have a common element s, s R a and s R b, so that b R a which is contrary to our hypothesis.

The set S can now be decomposed into equivalence classes T, T_1, T_2, ... such that every element of S belongs to one of these classes. Since these classes are mutually disjoint, we obtain the required partition of S.

Theorem 2:

If R is an equivalence relation in a non-empty set X, Then the quotient set X|R is a partition of x.

Proof :

Each $x \subset X$ must belong to some equivalence class. Also, the equivalence classes are pairwise disjoint, for if $z \in x|R \cap y|R$ then x R z, y R z, Since y R z $\Rightarrow$ z R y, x R z and z R y $\Rightarrow$ x R y, it follows that x|R and y|R must be identical. Hence two equivalence classes are either disjoint or identical. The set of equivalence classes is therefore a *partition.* Further, if x, y be any two members of the some set of this partition, they stand in a relation R to each other, showing that the partition induces the relation R.

Converse : If C be a partition of X, then the induced relation is an equivalence relation whose set of equivalence classes is X|C.

PRODUCT OF EQUIVALENCE RELATIONS

Let A, B be two sets and R, S be equivalence relations in A and B respectively.

The relation $R \times S$ in $A \times B$, defined by (x, y) $R \times S$ (z, u) $\Leftrightarrow$ x R z and y S u, is an equivalence relation in $A \times B$.

$R \times S$ is called the product of the relations R and S. Every equivalence class of $A \times B$ is of the form $P \times Q$ where P is an equivalence class of A

× B is of the form P × Q where P is an equivalence class of X and R and Q is an equivalence class of Y mod S.

TOTALLY ORDERED SET

The word 'partial-order' in the phrase 'partially ordered set' is inclined to emphasize that there may be some elements in a partially ordered set which are uncomparable.

If every two elements of a partially ordered set are comparable then the set A is called a *totally ordered set* or *linearly ordered set or chain, e.g.* the set N with usual order is totally ordered set.

SUBSET OF A PARTIALLY ORDERED SET

Let (A, R) be a partially ordered set and let $B \subset A$ be arbitrary. Then the partial order in A induces a partial order R, in B such that

$$aR'b : a, b \in B \Rightarrow aRb.$$

Then, we say that (B, R') is a subset of a partially ordered set (A, R). Generally the subset of a partially ordered set $(A, \leq)$ is denoted by $(B, \leq)$. *It is clear that any subset of a partially ordered set is a partially ordered set and similar is the case for a linearly ordered set.*

PARTIAL ORDER RELATIONS

A relation ≤ on a set A is called a *partial order relation iff* it is *reflexive, antisymmetric and transitive* and in this case the set A is called a *partial ordered set* and is denoted by the symbol $(A, \leq)$.

Remark : The relation in N defined by '$x < y$' is a partial order and is called *natural order or usual order in N.*

Comparable Elements : Two elements x, y in a partially ordered set $(A, \leq)$ are said to be *comparable* if either $x \leq y$ or $y \leq x$.

Uncomparable Elements : Two elements x, y in a partially ordered set $(A, \leq)$ are said to *uncomparable* if

$$x \leq y \text{ and } y \leq x \text{ both do not hold.}$$

Example : Let A = {1, 2, 3, 4, 5, 8, 16} Define a relation < is A by requiring $x < y$ iff $2x = y$.

The relation < in A is a partial order. Here the elements 1, 3 are not comparable. So, we write $1 \nless 3$.

Similarly, $1 \nless 5$, $5 \nless 1$, $3 \nless 8$, $8 \nless 3$.

The elements 4, 8 are comparable and so write $4 < 8$.

Similarly, $8 < 16$, $2 < 4$ etc.

UPPER AND LOWER BOUNDS

Let $(X, \leq)$ be a partially ordered set and let $(A, \leq)$ be a subset of $(X, \leq)$.

An element $a \in X$ is said to be an *upper bound of A if* $x \leq a \ \forall \ x \in A$.

An upper bound b of A is said to be a *least upper bound or supremum of A.* If $b \leq a$ for every upper bound a of A.

In brief, we write *l.u.b.* in place of 'least upper bound'. Similarly, we write *sup.* for the word supremum. The supremum of A is denoted by the symbol sup. (A).

An element $a \in X$ is called *lower bound of A* if

$$a \leq x \ \forall \ x \in A.$$

A lower bound b of A is called *greater lower bound or infimum of A,* if $a \leq b$ for every lower bound a of A.

In short, greatest lower bound is written as g.l.b.

Similarly, the infimum of A denoted by the symbol inf. (A).

Examples :

(i) Let $A = \{x \in Q; 2 < x^2 < 5\}$ be a set in the usual order. Then A contains an infinite number of lower bounds and upper bounds, but sup. (A) and inf. (A) do not exist.

(ii) Let $X = \{1, 2, 3, 4, 5, 6\}$ be a set in the natural order. Then $A = (2, 3, 4\}$ is a subset of the partially order set X.

Lower bounds of A are 2, 1.

Upper bounds of A are 4, 5, 6.

$\therefore$ sup (A) = 4, inf. (A) = 2.

SOLVED EXAMPLES

Example 1 : *Using mathematical induction prove that* $(1 + x)^n > 1 + nx$, *for* $n \geq 2$ *and* $x > -1$, $(\neq 0)$.

Solution : Let the given result be denoted by P(n).

When n = 1, we get

$$(1 + x) > 1 + x$$

Which is not true (this is given in the problem).

When n = 2, we get

$$(1 + x)^2 > 1 + 2x + x^2 > 1 + 2x$$

Which is true. Therefore, P(2) is true.

Let the result P(r) be true, *i.e.*,

$$(1 + x)^r > 1 + rx \qquad \text{...(1)}$$

We are to prove that P (r + 1) is true,

i.e., $(1 + x)^{r+1} > 1 + (r + 1)\, x$

Now $(1 + x)^{r+1} = (1 + x)^r (1 + x)$

$> (1 + rx)(1 + x)$ (using (1))

$= 1 + rx + x + rx^2$

$= 1 + (r + 1)\, x + rx^2$

$> 1 + (r + 1)\, x$

Therefore P (r + 1) is also true. Hence, P (n) is true for all $n \geq 2$, $n \in N$.

Example 2 : *By mathematical induction, prove that*

$$7^{2n} + (2^{3n-3})\, 3^{n-1} \text{ is divisible by } 25,\ n \in N.$$

Solution : Let the statement P(n) be defined as

P(n) = "$7^{2n} + (2^{3n-3})\, 3^{n-1}$ is divisible by 25"

When $n = 1$, we get

$P(1) = 7^2 + 1\,(1) = 50$

Which is divisible by 25,

Let the result P(r) be true.

That is $7^{2r} + (2^{3r-3})\, 3^{r-1}$ is divisible by 25

Let $7^{2r} + (2^{3r-3})\, 3^{r-1} = 25\, k,\ k \in N$ ....(1)

Now $P(r + 1) = 7^{2r+2} + (2^{3r})\, 3^r$

$= 7^{2r}(49) + (2^{3r-3} . 2^3)\, 3^{r-1} . 3$

$= 49\,(7^{2r}) + 24\,(2^{3r-3})\, 3^{r-1}$

$= (50 - 1)(7^{2r}) + (25 - 1)(2^{3r-3})3^{r-1}$

$= 50\,(7^{2r}) + 25\,(2^{3r-3})\, 3^{r-1} - [7^{2r} + (2^{3r-3})\, 3^{r-1}]$

$= 25\,[2\,(7^{2r}) + (2^{3r-3})\, 3^{r-1}] - 25\, k$

(from (1))

$= 25\,[2\,(7^{2r}) + (2^{3r-3})\, 3^{r-4} - k]$

= divisible by 25.

Therefore, P (r + 1) is also true. By mathematical induction, P(n) is true for all $n \in N$.

Example 3 : *Using mathematical induction prove that for every integer* $n \geq 1$, $\left(3^{2^n} - 1\right)$ is divisible by 2^{n+2} but not by 2^{n+3}.

Solution : Let P(n) be the statement P(n) : $3^{2^n} - 1$ is divisible by 2^{n+2} but not by 2^{n+3}.

For n = 1, we have $P(1) = 3^2 - 1 = 8 = 2^3$, which is divisible by 2^3 but not by 2^4.

Hence P(1) is true. Let the statement P(r) be true. That is

$$P(r) : 3^{2^r} - 1 \text{ is divisible by } 2^{r+2} \text{ but not by } 2^{r+3}$$

Therefore, we can write

$$P(r) : 3^{2^r} - 1 = k \,.\, 2^{r+2}$$

where k is odd (if k is even, then $k.2^{r+2}$ is divisible by 2^{r+3}).

Hence $3^{2^r} = 1 + k\, 2^{r+2}$, k odd.

Now, $\quad P(r + 1) : 3^{2^{r+1}} - 1 = 3^{(2^r,2)} - 1$

$$= \left(3^{2^r}\right)^2 - 1 = (1 + k \,.\, 2^{r+2})^2 - 1$$

$$= k^2 \,.\, 2^{2(r+2)} + 2k \,.\, 2\, 2^{r+2} + 1 - 1$$

$$= k^2 \,.\, 2^{2r+4} + k \,.\, 2^{r+3} = k^2 \,.\, 2^{r+3} \,.\, 2^{r+1} + k \,.\, 2^{r+3}$$

$$= k \,.\, 2^{2r+3}\, [1 + k \,.\, 2^{r+1}]$$

Which is divisible by 2^{r+3}, but not by 2^{r+4} as k is odd. Hence P(r + 1) is true. By mathematical induction, P(n) is true fro all $n \in N$.

Example 4 : *Let $p \geq 3$ be an integer and α, β be the roots of $x^2 - (p + 1)\, x + 1 = 0$. Using mathematical induction show that $\alpha^n + \beta^n$ is an integer.*

Solution : Since α, β are the roots of the equation

$x^2 - (p + 1)\, x + 1 = 0$, we have

$\alpha + \beta = p + 1$, and $\alpha\, \beta = 1$ ...(1)

Let $P(n) = \alpha^n + \beta^n$. We are to prove that P(n) is an integer.

For n = 1, we get

$P(1) = \alpha + \beta = p + 1$.

Which is an integer since p is an integer.

Hence, P(1) is true. Let P(r) be true.

That is, $P(r) = \alpha^r + \beta^r$ is an integer.

$$\begin{aligned}\text{Now,} \quad P(r+1) &= \alpha^{r+1} + \beta^{r+1} \\ &= \alpha(\alpha^r) + \beta(\beta^r) \\ &= \alpha(\alpha^r + \beta^r) - \alpha\beta^r + \beta(\alpha^r + \beta^r) - \beta\alpha^r \\ &= (\alpha + \beta)(\alpha^r + \beta^r) - \alpha\beta(\alpha^{r-1} + \beta^{r-1}) \\ &= (p+1)(\alpha^r + \beta^r) - (\alpha^{r-1} + \beta^{r-1}) \qquad ...(2)\end{aligned}$$

Since p is an integer, p + 1 is an integer and $\alpha^r + \beta^r$ is an integer. Therefore P (r + 1) is an integer, if $\alpha^{r-1} + \beta^{r-1}$ is an integer putting r = r – 1 in (2), we get

$$P(r) = \alpha^r + \beta^r = (p+1)(\alpha^{r-1} + \beta^{r-1}) - (\alpha^{r-2} + \beta^{r-2}) \qquad ...(3)$$

Again, since (p + 1) and $(\alpha^r + \beta^r)$ are integers, $\alpha^{r-1} + \beta^{r-1}$ is an integer, if $\alpha^{r-2} + \beta^{r-2}$ is an integer. Setting r = r–1 in (3), we find $\alpha^{r-2} + \beta^{r-2}$ is an integer this way, we get that $\alpha^2 + \beta^2$ is an integer if $\alpha + \beta$ is an integer. But $\alpha + \beta = p + 1$ is an integer.

Hence, $\alpha^2 + \beta^2$ is an integer, $\alpha^3 + \beta^3$ is an integer . . . . . $\alpha^{r-1} + \beta^{r-1}$ is an integer.

Hence, by mathematical induction, $\alpha^n + \beta^n$ is an integer.

Example 5: *Two finite sets have m and n elements. The total number of subsets of the first set is 56 more than the total number of subsets of the second set. Find the values of m and n.*

Solution: Let A and B be two sets having m and n elements respectively. Then,

Number of subsets of A = 2^m,

Number of subsets of B = 2^n,

It is given that $2^m - 2^n = 56$

$\Rightarrow \quad 2^n(2^{m-n} - 1) = 2^3(2^3 - 1)$

$\Rightarrow \quad n = 3$ and $m - n = 3$

$\Rightarrow \quad n = 3$ and $m = 6$.

Examples 6: *The necessary and sufficient condition for a set Y to be a subset of X is that $X \cup Y = X$.*

Solution:

Let $Y \subset X$, then

$$x \in Y \Rightarrow x \in X \quad \text{...(i)}$$

Now, $x \in X \cup Y \Rightarrow x \in X$ or $x \in Y$

$$\Rightarrow x \in X$$

$\therefore$ $X \cup Y \subset X$ ...(ii)

Also, we know that

$$X \subset X \cup Y \quad \text{...(iii)}$$

(ii) and (iii) $\Rightarrow X \cup Y = X$.

Conversely,

if $X \cup Y = X$, we have to prove that $Y \subset X$.

Now, $X \cup Y = X \Rightarrow X \cup Y \subset X$

and $X \subset X \cup Y$

$\Rightarrow$ $X \cup Y \subset X$

$\Rightarrow$ $X \subset X$ and $Y \subset X$

$\Rightarrow$ $Y \subset X$.

Example 7: *Prove that the operation* $*$ *on the set Z of all integers defined by* $a * b = a + b + 1 \ \forall a, b \in Z$ *satisfies the closure property the associativity and the commutativity. Find the identity element. What is the inverse of an integer a?*

Solution: For all integers a and b, $a \in b \in I \Rightarrow a + b + 1 \in I$ and therefore $a * b$ is clearly an integer. So, closure property is satisfied.

Also, $(a * b) * c = (a + b) * c \Rightarrow (a + b + 1)* c$

$\Rightarrow (a + b + 1) + c + 1 \Rightarrow a + b + c + 2$

and $a * (b * c) = a * (b + c + 1)$

$= a + (b + c + 1) + 1 = (a + b + c) + 2$

$\therefore$ $(a * b) * c = a * (b * c)$ " a, b, c $\in$ Z.

Thus, the associative law is satisfied.

Similarly, commutative law can be established.

Now, if e is the identity element in Z for $*$, then

$$a * e = a \Rightarrow a + e + 1 = a \Rightarrow e = -1.$$

So, -1 is the identity for $*$ in Z.

Also, $a * b = -1 \Rightarrow a + b + 1 = -1 \Rightarrow b - (2 + a)$.

So, the inverse of a is $-(2 + a)$.

Examples 8: *In a certain examination, the candidates can offer papers in English or Hindi or both the subjects. The number of candidates who appeared in the examination is 1000 of whom 650 appeared in English and 200 appeared in both English and Hindi. Find :*

(i) The number of candidates who offered paper in Hindi.

(ii) The number of candidates who offered paper in English only.

(iii) The number of candidates who offered paper in Hindi only.

Solution: Let A = The set of candidates who offered paper in English

B = The set of candidates who offered paper in Hindi.

Therefore, $n(A \cup B) = 1000$, $n(A) = 650$, $n(A \cap B) = 200$.

(i) We have

$$n(A \cup B) = n(A) + n(B) - n(A \cap B)$$

$$\Rightarrow \quad 1000 = 650 + n(B) - 200$$

$$\Rightarrow \quad n(B) = 550.$$

$\therefore$ Number of candidates who offered paper in Hindi = 550.

(ii) Now, the set of candidates who offered paper in English only is

$$A - B = \{\text{candidate: candidate} \in A \text{ and candidate} \notin B\}$$

Now, $n(A - B) = n(A) - n(A \cap B)$

$= 650 - 200 = 450.$

Number of candidates who offered paper in English only = 450.

(iii) Similarly, the set of candidates who offered paper in Hindi only is $B - A$.

Now, $n(B - A) = n(B) - n(A \cap B)$

$= 550 - 200 = 350.$

Example 9: *If $a \in N$ such that $aN = \{ax: x \in N\}$. Describe the set $3N \cap 7N$.*

Solution: We have

$$aN = \{ax\ ;\ x \in N\}$$

$$\therefore \quad 3N = [3x: x \in N\} = \{3, 6, 9, 12, \ldots,\}$$

and $\quad 7N = [7x: x \in N\} = \{7, 14, 21, 28, \ldots,\}$

Hence $\quad 3N \cap 7N = \{21, 42, \ldots\}$

$= \{21x: x \in N\} = 21N.$

Example 10: *For any natural number a, we define* $aN = \{ax: x \in N\}$. *If* $b, c, d \in N$ *such that* $bN \cap cN = dN$, *then prove that d is the l.c.m. of b and c.*

Solution: We have

$bN = \{bx: x \in N\}$ = The set of positive integral multiples of b.

$cN = \{cx: x \in N\}$ = The set of positive integral multiples of c.

$\therefore$ $bN \cap cN$ = The set of positive integral multiples of b and c both.

$\Rightarrow$ $bN \cap cN = \{kx: x \in N\}$, where k is the *l*.c.m. of b and c.

Hence, d = *l*.c.m of b and c.

Examples 11: *Let A, B, C be three sets then*

$$(A - B) \cap (A - C) = A - (B \cup C)$$

Solution: Let $x \in (A - B) \cap (A - C)$ then

$$\begin{aligned} x \in (A - B) \cap (A - C) &\Leftrightarrow x \in (A - B) \text{ and } x \in (A - C) \\ &\Leftrightarrow x \in A, x \notin B \text{ and } x \in A, x \notin C \\ &\Leftrightarrow x \subset A, x \notin B \text{ and } x \notin C \\ &\Leftrightarrow x \in A, x \notin B \cup C \\ &\Leftrightarrow x \in A - (B \cup C) \end{aligned}$$

Which implies that $(A - B) \cap (A - C) = A - (B \cup C)$

Example 12: *If A, B and C are any three sets, then prove that:*

(a) $A \times (B \cap C) = (A \times B) \cap (A \times C)$,

(b) $A \times (B \cup C) = (A \times B) \cup (A \times C)$.

Solution:

(a) We shall show that

$A \times (B \cap C) \subset (A \times B) \cap (A \times C)$

and $(A \times B) \cap (A \times C) \subset A \times (B \cap C)$

Firstly, let $(x, y) \in A \times (B \cap C)$, then

$$\begin{aligned} & (x, y) \in A \times (B \cap C) \\ \Rightarrow\ & x \in A \text{ and } y \in (B \cap C) \\ \Rightarrow\ & x \in A \text{ and } [y \in B \text{ and } y \in C] \\ \Rightarrow\ & [x \in A \text{ and } y \in B] \text{ and } [x \in A \text{ and } y \in C] \\ \Rightarrow\ & (x, y) \in (A \times B) \text{ and } (x, y) \in (A \times C) \\ \Rightarrow\ & (x, y) \in (A \times B) \cap (A \times C) \end{aligned}$$

Thus, $A \times (B \cap C) \subset (A \times B) \cap (A \times C)$ ...(1)

Again, let $(x, y) \in (A \times B) \cap (A \times C)$, then

$(x, y) \in (A \times B) \cap (A \times C)$

$\Rightarrow$ $(x, y) \in A \times B$ and $(x, y) \in (A \times C)$

$\Rightarrow$ $(x \in A, y \in B)$ and $(y \in A, y \in C)$

$\Rightarrow$ $x \in A, y \in B \cap C$

$\Rightarrow$ $(x, y) \in A \times B \cap C$

$\therefore$ $(A \times B) \cap (A \times C) \subset A (B \cap C)$ ...(2)

From (1) and (2)

$A \times (B \cap C) = (A \times B) \cap (A \times C)$

(b) Firstly, let $(x, y) \in A \times (B \cup C)$, then

$(x, y) \in A \times (B \cup C)$

$\Rightarrow$ $x \in A, y \in (B \cup C)$

$\Rightarrow$ $x \in A, [y \in B \text{ or } y \in C]$

$\Rightarrow$ $[x \in A, y \in B]$ or $[x \in A, y \in C]$

$\Rightarrow$ $(x, y) \in A \times B$ or $(x, y) \in A \times C$

$\Rightarrow$ $(x, y) \in (A \times B) \cup (A \times C)$.

Since, $(x, y) \in A \times (B \cup C)$

$\Rightarrow$ $(x, y) \in (A \times B) \cup (A \times C)$

$\therefore$ $A \times (B \cup C) \subset (A \times B) \cup (A \times C)$ ...(1)

Similarly, it can be proved that

$(x, y) \in (A \times B) \cup (A \times C)$

$\Rightarrow$ $(x, y) \in A \times (B \cup C)$,

$\therefore$ $(A \times B) \cup (A \times C) \subset A \times (B \cup C)$ ...(2)

From (1) and (2)

$A \times (B \cup C) = (A \times B) \cup (A \times C)$

Example 13: *If $A \subset B$ show that $A \times A \subset (A \times B) \cap (B \times A)$.*

Solution: Let $(x, y) \in A \times A$, then

$(x, y) \in A \times A$,

$\Rightarrow$ $x \in A, y \in A$

Since $A \subset B$ is given,

$x \in A \Rightarrow y \in B$.

$\therefore$ $(x, y) \in A \times A$

$\Rightarrow$ $x \in A, y \in B$

$\Rightarrow \quad (x, y) \in (A \times B)$ ...(1)

Also, $\quad (x, y) \in A \times A$

$\Rightarrow \quad x \in A, y \in A$

$\therefore \quad x \in B, y \in A$ $[\because A \subset B]$

$\Rightarrow \quad (x, y) \in B \times A$ ...(2)

Hence, from (1) and (2),

$(x, y) \in A \times A$

$\Rightarrow \quad (x, y) \in (A \times B)$ and $(x, y) \in B \times A$

$\Rightarrow \quad (x, y) \in (A \times B) \cap (B \times A)$

$\therefore \quad A \times A \subset (A \times B) \cap (B \times A).$

Example 14: *If $A \subset B$, prove that $A \times C \subset B \times C$ for any set C.*

Solution: Let (x, y) be any arbitrary element of $A \times C$. Then,

$(x, y) \in A \times C$

$\Rightarrow \quad x \in A$ and $y \in C$

$\Rightarrow \quad x \in B$ and $y \in C$ $[\because A \subset B]$

$\Rightarrow \quad (x, y) \in B \times C$

$\therefore \quad A \times C \subset B \times C.$

Example 15: *Prove, using mathematical induction that $10^{2n-1} + 1$ is divisible by 11 for all $n \in N$.*

Solution:

Let P(n) be the statement.

P(n): "$10^{2n-1} + 1$ is divisible by 11."

When $\quad n = 1$, we get $10^{2n-1} + 1 = 10 + 1 = 11$,

Which is divisible by 11. Hence, P(1) is true.

Let the result be true for P(r), *i.e.*, $10^{2r-1} + 1$ is divisible by 11. Therefore,

$$10^{2r-1} + 1 = 11k, k \in N.$$

or $\quad 10^{2r-1} = 11k - 1.$

We are now to prove that P(r + 1) is true.

Now, $\quad = P(r + 1): 10^{(2r+1)-1} - 1 = 10^{2r+1} + 1$

$= 10^{2r-1}.10^2 + 1 + (11k-1)\ 100 + 1 = 11(100k - 9)$

$=$ multiple of 11.

Hence, 10^{2r+1} is divisible by 11. That is, P(r + 1) is true. By the principle of mathematical induction P(n) is true for all $n \in N$.

Example 16:

Using mathematical induction prove that $(1 + x)^n > 1 + nx$, for $n \geq 2$ and $x > -1$, $(\neq 0)$.

Solution:

Let the given result be denoted by P(n).

When n = 1, we get

$$(1 + x) > 1 + x$$

Which is not true (this is given in the problem).

When n = 2, we get

$$(1 + x)^2 > 1 + 2x + x^2 > 1 + 2x$$

Which is true. Therefore, P(2) is true.

Let the result P(r) be true, *i.e.*,

$$(1 + x)^r > 1 + rx \quad \text{...(1)}$$

We are to prove that P (r + 1) is true,

i.e., $(1 + x)^{r+1} > 1 + (r + 1)x$

Now $(1 + x)^{r+1} = (1 + x)^r (1 + x)$

$$> (1 + rx)(1 + x) \quad \text{(using (1))}$$

$$= 1 + rx + x + rx^2$$

$$= 1 + (r + 1)x + rx^2$$

$$> 1 + (r + 1)x$$

Therefore P (r + 1) is also true. Hence, P (n) is true for all $n \geq 2$, $n \in N$.

Examples 17: *Prove that $(A')' = A$.*

Solution: We have to prove $(A')' \subset A$ and $A \subset (A')'$.

Let $x \in (A')'$ then

$$x \in (A')' \Leftrightarrow x \notin A'$$

$$\Leftrightarrow x \in A$$

Thus $x \in (A')' \Leftrightarrow x \in A$

$$\Rightarrow (A')' \subset A \text{ and } (A')' \supset A$$

$$\Rightarrow (A')' = A.$$

Example 18: *Using mathematical induction prove that for every integer* $n \geq 1$, $\left(3^{2^n} - 1\right)$ is divisible by 2^{n+2} but not by 2^{n+3}.

Solution: Let P(n) be the statement P(n): $3^{2^n} - 1$ is divisible by 2^{n+2} but not by 2^{n+3}. For n = 1, we have $P(1) = 3^2 - 1 = 8 = 2^3$, which is divisible by 2^3 but not by 2^4.

Hence P(1) is true. Let the statement P(r) be true. That is

$$P(r): 3^{2^r} - 1 \text{ is divisible by } 2^{r+2} \text{ but not by } 2^{r+3}$$

Therefore, we can write

$$P(r): 3^{2^r} - 1 = k \cdot 2^{r+2}$$

where k is odd (if k is even, then $k.2^{r+2}$ is divisible by 2^{r+3}).

Hence $3^{2^r} = 1 + k\, 2^{r+2}$, k odd.

Now, $\quad P(r + 1): 3^{2^{r+1}} - 1 = 3^{(2^r, 2)} - 1$

$$= \left(3^{2^r}\right)^2 - 1 = (1 + k \cdot 2^{r+2})^2 - 1$$

$$= k^2 \cdot 2^{2(r+2)} + 2k \cdot 2\, 2^{r+2} + 1 - 1$$

$$= k^2 \cdot 2^{2r+4} + k \cdot 2^{r+3} = k^2 \cdot 2^{r+3} \cdot 2^{r+1} + k \cdot 2^{r+3}$$

$$= k \cdot 2^{2r+3}\, [1 + k \cdot 2^{r+1}]$$

Which is divisible by 2^{r+3}, but not by 2^{r+4} as k is odd. Hence P(r + 1) is true. By mathematical induction, P(n) is true fro all $n \in N$.

Example 19: *Let* $p \geq 3$ *be an integer and* α, β *be the roots of* $x^2 - (p + 1)x + 1 = 0.]$

Solution: Since α, β are the roots of the equation $x^2 - (p + 1)x + 1 = 0$, we have

$$\alpha + \beta = p + 1, \text{ and } \alpha\beta = 1 \qquad ...(1)$$

Let $P(n) = \alpha^n + \beta^n$. We are to prove that P(n) is an integer.

For n = 1, we get

$P(1) = \alpha + \beta = p + 1.$

Which is an integer since p is an integer.

Hence, P(1) is true. Let P(r) be true.

That is, $P(r) = \alpha^r + \beta^r$ is an integer.

Now, $\quad P(r + 1) = \alpha^{r+1} + \beta^{r+1}$

$$= \alpha\,(\alpha^r) + \beta(\beta^r)$$
$$= \alpha\,(\alpha^r + \beta^r) - \alpha\beta^r + \beta\,(\alpha^r + \beta^r) - \beta\,\alpha^r$$
$$= (\alpha + \beta)\,(\alpha^r + \beta^r) - \alpha\beta(\alpha^{r-1} + \beta^{r-1}$$
$$= (p + 1)\,(\alpha^r + \beta^r) - (\alpha^{r-1} + \beta^{r-1}). \qquad ...(2)$$

Since p is an integer, p + 1 is an integer and $\alpha^r + \beta^r$ is an integer. Therefore P (r + 1) is an integer, if $\alpha^{r-1} + \beta^{r-1}$ is an integer putting r = r – 1 in (2), we get

$$P\,(r) = \alpha^r + \beta^r = (p + 1)\,(\alpha^{r-1} + \beta^{r-1}) - (\alpha^{r-2} + \beta^{r-2}) \qquad ...(3)$$

Again, since (p + 1) and $(\alpha^r + \beta^r)$ are integers, $\alpha^{r-1} + \beta^{r-1}$ is an integer, if $\alpha^{r-2} + \beta^{r-2}$ is an integer. Setting r = r–1 in (3), we find $\alpha^{r-2} + \beta^{r-2}$ is an integer this way, we get that $\alpha^2 + \beta^2$ is an integer if $\alpha + \beta$ is an integer. But $\alpha + \beta = p + 1$ is an integer.

Hence, $\alpha^2 + \beta^2$ is an integer, $\alpha^3 + \beta^3$ is an integer . . . . . $\alpha^{r-1} + \beta^{r-1}$ is an integer.

Hence, by mathematical induction, $\alpha^n + \beta^n$ is an integer.

Examples 20: *Prove the following*

(a) $A \,\Delta\, \phi = A$,

(b) $A \,\Delta\, A = \phi$,

(b) $A \,\Delta\, B = \phi \Leftrightarrow A = B$.

Solution:

(a) $A \,\Delta\, \phi = (A - \phi) \cup (\phi - A)$

$$= A \cup \phi \qquad (\because A - \phi = A,\ \phi - A = \phi)$$
$$= A.$$

(b) $A \,\Delta\, A = (A - A) \cup (A - A)$

$$= \phi \cup \phi = \phi$$

(c) $A \,\Delta\, B = \phi \;\Rightarrow (A - B) \cup (B - A) = \phi$

$$\Rightarrow A - B = \phi \text{ and } B - A = \phi$$
$$\Rightarrow A = B$$

and $A = B \Rightarrow A - B = \phi$ and $B - A = \phi$

$$\Rightarrow (A - B) \cup (B - A) = \phi$$
$$\Rightarrow A \,\Delta\, B = \phi$$

Hence, $A \,\Delta\, B = \phi \Rightarrow A = B$.

Examples 21: *If* *A = set of all students,*

B = set of all students offering mathematics,

C = set of all women,

D = set of industrious person,

E = set of all first class students.

Write the following statement symbolically. Some industrious women students are the students of mathematics but they are not first class".

Solution: Set of women students = $A \cap C$. Set of industrious women students

$$= A \cap C \cap D$$

Set of industrious women students offering mathematics

$$= A \cap C \cap D \cap B.$$

Set of all those industrious women students offering mathematics who are not first class = $A \cap C \cap D \cap B - E$. According to the statement there are some students of this type. Hence,

$$A \cap C \cap D \cap B - E \neq \phi.$$

Example 22: *By mathematical induction, prove that $7^{2n} + (2^{3n-3})3^{n-1}$ is divisible by 25, $n \in N$.*

Solution: Let the statement P(n) be defined as

$P(n) =$ "$7^{2n} + (2^{3n-3})\, 3^{n-1}$ is divisible by 25"

When $n = 1$, we get

$P(1) = 7^2 + 1\,(1) = 50$

Which is divisible by 25,

Let the result P(r) be true.

That is $7^{2r} + (2^{3r-3})\, 3^{r-1}$ is divisible by 25

Let $7^{2r} + (2^{3r-3})\, 3^{r-1} = 25\,k,\ k \in N$ ....(1)

Now $P(r+1) = 7^{2r+2} + (2^{3r})\, 3^{r}$

$= 7^{2r}\,(49) + (2^{3r-3}\,.\,2^3)\, 3^{r-1}\,.\,3$

$= 49\,(7^{2r}) + 24\,(2^{3r-3})\, 3^{r-1}$

$= (50-1)\,(7^{2r}) + (25-1)\,(2^{3r-3})3^{r-1}$

$= 50\,(7^{2r}) + 25\,(2^{3r-3})\, 3^{r-1} - [7^{2r} + (2^{3r-3})\, 3^{r-1}]$

$= 25\,[2\,(7^{2r}) + (2^{3r-3})\, 3^{r-1}] - 25\,k$ (from (1))

$= 25\,[2\,(7^{2r}) + (2^{3r-3})\, 3^{r-4} - k]$

$= \text{divisible by } 25.$

Therefore, P (r + 1) is also true. By mathematical induction, P(n) is true for all $n \in N$.

Example 23 : *Show that* $n\{P[P.(P(\phi))]\} = 4$

Solution : We have $P(\phi) = \{\phi\}$

$$\therefore \quad P(P(\phi)) = \{\phi, \{\phi\}\}$$

$$\Rightarrow \quad P[P(P(\phi))] = \{\phi, \{\phi\}\}, [\{\phi\}] \{\phi, \{\phi\}\}\}$$

$$\therefore \quad n\{P[P.(P(\phi))]\} = 4.$$

Example 24 : *If* $A = \{a, \{b\}\}$, *find* $P(A)$.

Solution : Let $B = \{b\}$. Then $A = \{a, B\}$

$$\therefore \quad P(A) = \{\phi, \{a\}, \{B\}, \{a, B\}\}$$

$$= \{\phi, \{a\}, \{\{b\}\}, \{\{a\}, \{b\}\}\}$$

Example 25: *Set A has three elements and set B has six elements. What can be the minimum number of elements in the set* $A \cup B$? *Find also the maximum number of elements in* $A \cup B$.

Solution: We have $n(A) = 3$, and $n(B) = 6$. Therefore, the maximum number of elements in $A \cap B$ is 3, since

$$n(A \cup B) = n(A) + n(B) - n(A \cap B)$$

the minimum number of elements in $A \cup B$ is given by

$$\text{minimum } [n(A \cup B)] = n(A) + n(B) - \text{max. } [n(A \cap B)]$$

$$= 3 + 6 - 3 = 6$$

Therefore, the minimum number of elements in $A \cup B = 6$.

The number of elements in $A \cup B$ is maximum when A and B are disjoint. In this case

$$n(A \cup B) = n(A) + n(B) = 3 + 6 = 9$$

The maximum number of elements in $A \cup B$ is 9.

Examples 26 : *The necessary and sufficient condition for a set Y to be a subset of X is that* $X \cup Y = X$.

Solution : Let $\quad Y \subset X$, then

$$x \in Y \Rightarrow x \in X \quad \text{...(i)}$$

Now, $\quad x \in X \cup Y \Rightarrow x \in X \text{ or } x \in Y$

$$\Rightarrow x \in X$$

$\therefore \quad X \cup Y \subset X$ ...(ii)

Also, we know that

$X \subset X \cup Y$ ...(iii)

(ii) and (iii) $\Rightarrow X \cup Y = X$.

Conversely,

if $\quad X \cup Y = X$, we have to prove that $Y \subset X$.

Now, $\quad X \cup Y = X \Rightarrow X \cup Y \subset X$

and $\quad X \subset X \cup Y$

$\Rightarrow \quad X \cup Y \subset X$

$\Rightarrow \quad X \subset X$ and $Y \subset X$

$\Rightarrow \quad Y \subset X$.

Examples 27 : *Prove that $(A')' = A$.*

Solution : We have to prove $(A')' \subset A$ and $A \subset (A')'$.

Let $\quad x \in (A')'$ then

$x \in (A')' \Leftrightarrow x \notin A'$

$\Leftrightarrow x \in A$

Thus $\quad x \in (A')' \Leftrightarrow x \in A$

$\Rightarrow (A')' \subset A$ and $(A')' \supset A$

$\Rightarrow (A')' = A$.

Examples 28 : *Let A, B, C be three sets then*

$$(A - B) \cap (A - C) = A - (B \cup C)$$

Solution : Let $x \in (A - B) \cap (A - C)$ then

$x \in (A - B) \cap (A - C) \Leftrightarrow x \in (A - B)$ and $x \in (A - C)$

$\Leftrightarrow x \in A, x \notin B$ and $x \in A, x \notin C$

$\Leftrightarrow x \in A, x \notin B$ and $x \notin C$

$\Leftrightarrow x \in A, x \notin B \cup C$

$\Leftrightarrow x \in A - (B \cup C)$

Which implies that $(A - B) \cap (A - C) = A - (B \cup C)$

Examples 29 : *Prove the following*

(a) $A \, \Delta \, \phi = A$,

(b) $A \Delta A = \phi$,

(b) $A \Delta B = \phi \Leftrightarrow A = B$.

Solution : (a) $A \Delta \phi = (A - \phi) \cup (\phi - A)$

$= A \cup \phi \quad (\because A - \phi = A, \phi - A = \phi)$

$= A$.

(b) $A \Delta A = (A - A) \cup (A - A)$

$= \phi \cup \phi = \phi$

(c) $A \Delta B = \phi \Rightarrow (A - B) \cup (B - A) = \phi$

$\Rightarrow A - B = \phi$ and $B - A = \phi$

$\Rightarrow A = B$

and $A = B \Rightarrow A - B = \phi$ and $B - A = \phi$

$\Rightarrow (A - B) \cup (B - A) = \phi$

$\Rightarrow A \Delta B = \phi$

Hence, $A \Delta B = \phi \Rightarrow A = B$.

Examples 30 : *If* A = *set of all students,*

B = *set of all students offering mathematics,*

C = *set of all women,*

D = *set of industrious person,*

E = *set of all first class students.*

Write the following statement symbolically. Some industrious women students are the students of mathematics but they are not first class".

Solution : Set of women students = $A \cap C$. Set of industrious women students

$$= A \cap C \cap D$$

Set of industrious women students offering mathematics

$$= A \cap C \cap D \cap B.$$

Set of all those industrious women students offering mathematics who are not first class = $A \cap C \cap D \cap B - E$. According to the statement there are some students of this type. Hence,

$$A \cap C \cap D \cap B - E \neq \phi.$$

Example 31 : *Set A has three elements and set B has six elements. What can be the minimum number of elements in the set $A \cup B$? Find also the maximum number of elements in $A \cup B$.*

Solution : We have $n(A) = 3$, and $n(B) = 6$. Therefore, the maximum number of elements in $A \cap B$ is 3, since

$$n(A \cup B) = n(A) + n(B) - n(A \cap B)$$

the minimum number of elements in $A \cup B$ is given by

$$\text{minimum } [n(A \cup B)] = n(A) + n(B) - \text{max. } [n(A \cap B)]$$
$$= 3 + 6 - 3 = 6$$

Therefore, the minimum number of elements in $A \cup B = 6$.

The number of elements in $A \cup B$ is maximum when A and B are disjoint. In this case

$$n(A \cup B) = n(A) + n(B) = 3 + 6 = 9$$

The maximum number of elements in $A \cup B$ is 9.

Examples 32 : *In a certain examination, the candidates can offer papers in English or Hindi or both the subjects. The number of candidates who appeared in the examination is 1000 of whom 650 appeared in English and 200 appeared in both English and Hindi. Find,*

(i) the number of candidates who offered paper in Hindi.

(ii) the number of candidates who offered paper in English only.

(iii) the number of candidates who offered paper in Hindi only.

Solution : Let A = The set of candidates who offered paper in English

B = The set of candidates who offered paper in Hindi.

Therefore, $n(A \cup B) = 1000$, $n(A) = 650$, $n(A \cap B) = 200$.

(i) We have $n(A \cup B) = n(A) + n(B) - n(A \cap B)$

$$\Rightarrow \quad 1000 = 650 + n(B) - 200$$

$$\Rightarrow \quad n(B) = 550$$

$\therefore$ Number of candidates who offered paper in Hindi = 550.

(ii) Now, the set of candidates who offered paper in English only is

$$A - B = \{\text{candidate : candidate} \in A \text{ and candidate} \notin B\}$$

Now, $n(A - B) = n(A) - n(A \cap B)$

$$= 650 - 200 = 450.$$

Number of candidates who offered paper in English only = 450.

(iii) Similarly, the set of candidates who offered paper in Hindi only is $B - A$.

Now, $n(B - A) = n(B) - n(A \cap B)$

$= 550 - 200 = 350.$

***Example 33** : If $a \in N$ such that $aN = \{ax : x \in N\}$. Describe the set $3N \cap 7N$.*

Solution : We have $aN = \{ax ; x \in N\}$

$\therefore \quad 3N = [3x : x \in N\} = \{3, 6, 9, 12, \ldots,\}$

and $\quad 7N = [7x : x \in N\} = \{7, 14, 21, 28, \ldots,\}$

Hence $\quad 3N \cap 7N = \{21, 42, \ldots\}$

$= \{21x : x \in N\} = 21N.$

***Example 34** : For any natural number a, we define $aN = \{ax : x \in N\}$. If $b, c, d \in N$ such that $bN \cap cN = dN$, then prove that d is the l.c.m. of b and c.*

Solution : We have

$bN = \{bx : x \in N\}$ = The set of positive integral multiples of b.

$cN = \{cx : x \in N\}$ = The set of positive integral multiples of c.

$\therefore \quad bN \cap cN$ = The set of positive integral multiples of b and c both.

$\Rightarrow \quad bN \cap cN = \{kx : x \in N\}$, where k is the *l*.c.m. of b and c.

Hence, $\quad d = $ *l*.c.m of b and c.

***Example 35** : Two finite sets have m and n elements. The total number of subsets of the first set is 56 more than the total number of subsets of the second set. Find the values of m and n.*

Solution : Let A and B be two sets having m and n elements respectively.

Then,

Number of subsets of A $= 2^m$,

Number of subsets of B $= 2^n$,

It is given that $2^m - 2^n = 56$

$\Rightarrow \quad 2^n(2^{m-n} - 1) = 2^3(2^3 - 1)$

$\Rightarrow \quad n = 3$ and $m - n = 3$

$\Rightarrow \quad n = 3$ and $m = 6$.

***Example 36** : If $X = \{4^n - 3n - 1 : n \in N\}$*

and $\quad Y = \{9(n - 1) : n \in N\}$, prove that $X \subset Y$.

Solution : Let $\quad x_n = 4^n - 3n : n \in N$. Then $x_1 = 4 - 3 - 1 = 0$.

And for $n \geq 2$, we have $x_n = 4^n - 3n - 1$

$$= (1 + 3)^n - 3n - 1$$

$$= {}^nc_0 + {}^nc_1.3 + {}^nc_2 + 3^2 + \ldots + {}^nc_n.3^n - 3n - 1$$

[Using Binomial Theorem]

$$= 1 + 3n + {}^nc_2.3^2 + {}^nc_3.3^3 + \ldots + {}^nc_n.3^n - 3n - 1$$

$[\because {}^nc_n = 1, {}^nc_1 = n]$

$$= 3^2 \{{}^nc_2 + {}^nc_3.3 + {}^nc_4.3^2 + \ldots + {}^nc_n.3^{n-2}\}$$

$$= 9 \{{}^nc_2 + {}^nc_3.3 + {}^nc_4.3^2 + \ldots + {}^nc_n.3^{n-2}\}$$

Thus, x_n is some positive integral multiple of 9 for $n \geq 2$.

Hence, X consists of all positive integral multiples of 9 of the form

$$9 [{}^nc_2 + 3.{}^nc_3 + 3^2.{}^nc_4 + \ldots + 3^{n-2}.{}^nc_n]$$

together with zero.

Now, $Y = \{9(n - 1) : n \in N\}$. This shows that Y consists of all integral multiples of 9 together with 0.

In fact, we have $X = \{0, 9, 54, 243, \ldots\}$

and $Y = \{0, 9, 18, 27, 36, 54, \ldots\}$

Thus $X \subset Y$.

Example 37 : *Suppose $A_1, A_2, \ldots, A_{30}$ are thirty sets each with five elements and $B_1, B_2, \ldots, B_n$ are n sets each with three element.*

Let $\bigcup_{i=1}^{30} A_i = \bigcup_{j=1}^{n} B_j = S.$

Assume that each element of S belongs to exactly ten of the A_i's and exactly 9 of B_j's. Find n.

Solution : Since each A_i has 5 elements and each element of S belongs to exactly 10 of A_i's.

$$\therefore \quad S = \bigcup_{i=1}^{30} B_j \Rightarrow n(S) = \frac{1}{10}\sum_{i=1}^{30} n(A_i)$$

$$= \frac{1}{10}(5 \times 30) = 15 \qquad \ldots(1)$$

Again each B_j has 3 elements and each element of S belongs to exactly 9 of B_j's.

$$\therefore \quad S = \bigcup_{j=1}^{n} B_j \Rightarrow n(S) = \frac{1}{9}\sum_{i=1}^{30} n(B_j)$$

$$= \frac{1}{9}(3n) = \frac{n}{3} \qquad \ldots(2)$$

From (i) and (2), we get

$$15 = \frac{n}{3} \Rightarrow n = 45.$$

Example 38 : *In town of 10,000 families it was found that 40% families buy newspaper A, 20% families newspaper B and 10% families buy newspaper C. 5% families buy A and B, 3% buy B and C and 4% buy A and C. If 2% families buy all the three newspapers, find the number of families which buy (i) A only (ii) B only (iii) None of A, B and C.*

Solution : Let P, Q and R be the set of families buying newspaper A, B and C respectively.

Let U be the universal set.

$n(P) = 40\%$ of $10,000 = 4000$,

$n(Q) = 20\%$ of $10,000 = 2000$,

$n(R) = 10\%$ of $10,000 = 1000$,

$n(P \cap Q) = 5\%$ of $10,000 = 500$,

$n(Q \cap R) = 3\%$ of $10,000 = 300$,

$n(R \cap P) = 4\%$ of $10,000 = 400$,

$n(P \cap Q \cap R) = 2\%$ of $10,000 = 200$,

and $n(U) = 10,000$.

(i) Required number $= n(P \cap Q' \cap R')$

$= n(P \cap (Q' \cap R)')$

$= n(P) - n[(P \cap (Q \cup R)]$ $[\because n(A \cap B)' = n(A - n(A \cap B)]$

$= n(P) - n[(P \cap Q) \cup (P \cap R)]$

$= n(P) - [n(P \cap Q) + n(P \cap R) - n\{(P \cap Q) \cap (P \cap R)\}]$

$= n(P) - [n(P \cap Q) + n(P \cap R) - n\{(P \cap Q \cap R)]$

$= 4000 - [500 + 400 - 200]$

$= 3300.$

(ii) Required number $= n(P' \cap Q \cap R')$

$= n(Q \cap (P' \cap R')$

$= n(Q \cap (P \cup R)')$

$= n(Q) - n[(Q \cap (P \cup R)]$ $[\because n(A \cap B)' = n(A) - n(A \cap B)]$

$= n(Q) - n[(Q \cap P) \cup (Q \cap R)]$

$= n(Q) - [n(Q \cap P) + n(Q \cap R) - n\{(Q \cap P) \cap (Q \cap R)\}]$

$= n(Q) - [n(P \cap Q) + n(Q \cap R) - n\{(P \cap Q \cap R)]$

$= 2000 - [500 + 300 - 200]$

$= 1400.$

(iii) Required number $= n(P' \cap Q' \cap R')$

$= n[(P \cup Q \cup R)']$

$= n(U) - n(P \cup Q \cup R)$

$= n(U) - n[n(P) + n(Q) + n(R) - n(P \cap Q) - n(Q \cap R)$
$- n(R \cap P) + n(P \cap Q \cap R)$

$= 10000 - [4000 + 2000 + 1000 - 500 - 300 - 400 + 200]$

$= 4000.$

Example 39 : *Prove that if $A \subset B$ and $C \subset D$, then $A \times C \subset B \times D$*

Solution : Let $(x, y) \in A \times C$, then

$(x, y) \in A \times C$

$\Rightarrow$ $x \in B$ and $y \in D$ $[\because A \subset B, C \subset D]$

$\Rightarrow$ $(x, y) \in B \times D$

$\therefore$ $A \times C \subset B \times D.$

Example 40: *Let A be the set of all triangles in a plane. Let R be the relation in A defined as xRy iff x is congruent to y, $x \in A$, $y \in A$. Then R is an equivalence relation.*

Solution: Hence, we observe that

1. xRx, for every $x \in A$, since every triangle is congruent to itself. Thus R is reflexive.
2. xRy $\Rightarrow$ yRx, since if triangle x is congruent to triangle y then y is congruent to x. Thus r is symmetric.
3. xRy and yRx $\Rightarrow$ xRz, since if triangle x is congruent to y and triangle y is congruent to z, the triangle x is congruent to z. Thus R is transitive.

Since R is reflexive, symmetric and transitive, therefore R is an equivalence relation.

Note: In a similar way we can prove that the relation of similarity in the set of all triangles in a plane is an equivalence relation.

Example 41 : *Prove that the relation defined by 'is perpendicular to' in the set of straight lines in a plane is symmetric but neither reflexive nor transitive.*

Solution : Let R stands for "is perpendicular to". Let L be the set of straight lines in a plane.

1. *Reflexivity :* Since no straight line is perpendicular to itself, hence relation is not reflexive, *i.e.*,

   a R a for a ∈ L

2. *Symmetry :* If straight line a is ⊥ to straight line b then b is also ⊥ to a *i.e.*,

   $$a \, R \, b \Rightarrow b \, R \, a \text{ for } a, b \in L.$$

   Hence relation is symmetric.

3. *Transitivity :* If for a, b, c, ∈ L, a ⊥ b and b ⊥ c (it may be parallel).

   Hence a R b, b R c does not implied a R c ∀ a, b, c ∈ L.

   Hence relation is not transitive.

Example 42 : *If I be the set of integers and if R be defined over I by "a R b iff a – b is an even integer" where a, b ∈ I, then show that the relation R is an equivalence relation.*

Solution :

(i) Reflexivity : Since O = a – a is even.

Hence a R a ∀ a ∈ I

Therefore, R is reflexive over I.

(ii) *Symmetry :* Since a – b is even then b – a = – (a – b) is also even and hence a R b

⇒ b R a.

Therefore, R is symmetric in I.

(iii) *Transitivity :* a R b and b R c

⇒ (a – b) and (b – c) are even numbers

⇒ (a – b) + (b – c) is even

⇒ (a – c) is even

∴ a R b and b R c

⇒ a R c.

Therefore, R is transitive in I.

Hence R is an equivalence relation.

Example 43 : *If I the set of integers then show that relation R in I such that a R b iff a – b is divisible by m (positive integer) is an equivalence relation.*

Solution : (i) *Reflexivity* : For each a ∈ I, a = a = 0 which is divisible by m, so a R a Therefore, R is reflexive.

(ii) *Symmetry* : Let a R b for a, b, ∈ I,

i.e., a – b is divisible by m. Then as a – b = – (b – a)

(b – a) is also divisible by m.

Hence a R b ⇒ b R a so that R is symmetric.

(iii) *Transitivity* : For a, b, c ∈ I, Let a R b, b R c, *i.e.*, let (a – b), (b – c) be divisible by m. Then as

(a – b) + (b – c) = a – c,

a – c is also divisible by m.

Hence, a R b, b R c ⇒ a R c, *i.e.*, R is transitive, showing that R is an equivalence relation.

Example 44: *Prove that the relation R in the set of integers I defined by a R b if a and b are both odd is symmetric and transitive but not reflexive.*

Solution:

1. *R is not Reflexive:* We have 2 ∈ I. But 2 is not R-related to 2, since 2 and 2 are not both odd. Therefore R is not reflexive.
2. *R is Symmetric:* Suppose we have a R b. Then a and b are both odd. Now if a and b are both odd then b and a are both odd *i.e.,* b R a. Thus a R b

   ⇒ b R a and therefore R is symmetric.
3. *R is Transitive:* Suppose we have a R b and b R c. Then a and b are both odd as well as b and c are both odd. It implies that a and c are both odd *i.e.,* a R c. Thus a R b and b R c

   ⇒ a R c. Therefore R is transitive.

Example 45: *Show with the help of an example that the union of two transitive relations is not necessarily a transitive relation.*

Solution: Let R_1 = {(1, 2), (2, 1), (1, 1), (2, 2)}. R_2{(2, 3), (3, 2), (2, 2), (3, 3)} be two relations on the set S = {1, 2, 3}. Obviously both R_1 and R_2 are transitive relations. Now $R_1 \cup R_2$ = {(1, 2), (2, 1), (1, 1), (2, 2), (2, 3), (3, 2), (3, 2), (2, 2), (3, 3)}.

This is not a transitive relation because $(1, 2) \in R_1 \cup R_2$ and $(2, 3) \in R_1 \cup R_2$ while $(1, 3) \notin R_1 \cup R_2$.

Example 46: *Prove that the relation "congruence modulo m" is an equivalence relation in the set of integers.*

Solution: Let I be the set of integers. If m is any given positive integer and a, b $\in$ I, then we say that $a \equiv b \pmod{m}$ if $m \mid (a - b)$ *i.e.,* if m is a divisor of a – b. We shall prove that this defines an equivalence relation in the set I.

Reflexivity: Let a be any integer. Then a – a = 0 and $m \mid 0$ because we can write 0 = m. 0. Thus $a \equiv a \pmod{m}$ $\forall$ a $\in$ I. Therefore the relation is reflexive.

Symmetry: Let a, b $\in$ I be such that $a \equiv b \pmod{m}$.

Then $a \equiv b \pmod{m}$

$\Rightarrow m \mid (a - b)$

$\Rightarrow a - b = km$, where k is some integer

$\Rightarrow b - a = (-k)\, m$, where –k is also an integer

$\Rightarrow m \mid (b - a)$

$\Rightarrow b \equiv a \pmod{m}$.

Thus, $a \equiv b \pmod{m}$

$\Rightarrow b \equiv a \pmod{m}$. Therefore the relation is symmetric.

Transitivity: Let a, b, c $\in$ I be such that

$a \equiv b \pmod{m}$, $b \equiv a \pmod{m}$. Then we have

$m \mid (a - b)$ and $m \mid (b - c)$

$\Rightarrow a - b = k_1 m$ and $b - c = k_2 m$, where k_1, k_2 are some integers

$\Rightarrow (a - b) + (b - c) = k_1 m + k_2 m$

$\Rightarrow a - c = (k_1 + k_2)m$, where $k_1 + k_2$ is also an integer

$\Rightarrow m \mid (a - c) \Rightarrow a \equiv c \pmod{m}$.

Thus $a \equiv b \pmod{m}$ and $b \equiv c \pmod{m}$

$\Rightarrow a \equiv c \pmod{m}$. Therefore the relation is transitive.

Since the relation of "congruence modulo m" on the set of integers is reflexive, symmetric and transitive, therefore it is an equivalence relation.

Example 47: *What is the difference between function and relation?*

Let X = {1, 2, 3, 4}

and Y = {1, 5, 9, 11, 15, 16}.

Determine which of the following sets are :

(i) relation (ii) function (iii) neither, of X to Y.

(a) $f_1 = \{(x, y) \mid y = x^2, x \in X, y \in Y\}$

(b) $f_2 = \{(1, 1), (2, 11), (3, 1), (4, 15)\}$

(c) $f_3 = \{(1, 5), (2, 9), (3, 1), (4, 5), (2, 11)\}$

(d) $f_4 = \{(1, 1), (2, 7), (3, 5)\}$.

Solution: For difference between function and relation refer

(a) We have $f_1 = \{(1, 1), (3, 9), (4, 16)\}$.

Hence f_1 is a relation of X to Y because $f_1 \subseteq X \times Y$. But f_1 is not a function of X to Y because $2 \in X$ and $2^3 = 4$ which $\notin$ Y. Thus under f_1 the image of 2 belonging to X does not exist.

(b) Hence f_2 is a relation of X to Y because $f_2 \subseteq X \times Y$. Also f_2 is a function of X to Y because under f_2 each element of X has a unique image in Y.

(c) Hence, f_3 a relation of X to Y because $f_3 \subseteq X \times Y$. But f_3 is not a function of X to Y because under f_3 the image of $2 \in X$ is not unique. Note the $f_3(2) = 9$ and also $f_3(2) = 11$.

(d) Here f_4 is not a relation of X to Y because $f_4 \not\subseteq X \times Y$. Note that $(2, 7) \in f_4$ but $(2, 7) \notin X \times Y$ because $7 \notin$ Y. Also f_4 is not a function of X to Y.

Example 48: *Which of the following relations in the set of real numbers are equivalence relations ?*

1. $a\ R\ b$ *iff* $|a| \le b$,
2. $a\ R\ b$ *iff* $|a| \ge |b|$
3. $a\ R\ b$ *iff* $|a| \ne |b|$.

Solution: Let E denote the set of real numbers.

1. The given relation R on the set E is not reflexive.

   For example $-6 \in E$. We have $|-6| = 6$ and the statement $6 \le -6$, is not true. Thus -6 is not R-related to -6 because the statement $|-6| \le -6$ is not true.

   Thus, there exists elements in E which are not R-related to themselves. Therefore the relation r is not reflexive. Hence it is not an equivalence relation.

2. The given relation R on the set E is not reflexive. For example $3 \in E$. We have $|3| = 3$. The statement $|3| > |3|$ is false. Thus

3 is not R-related to 3 because the statement | 3 | > | 3 | is not true. Thus, there exist element in E which are not R-related to themselves. Therefore the relation R on the set E is not reflexive. Hence it is not an equivalence relation.

3. The given relation R on the set E is not reflexive. For example $4 \in E$. The statement $| 4 | \neq | 4 |$ is false and so 4 is no R-related to itself. Since the relation R on the set E is not reflexive, therefore it cannot be an equivalence relation.

Example 49: *Show that the relation R in the set of non-zero integers I_0 defined by a R b if $a^b = b^a$, $(a, b, \in I_0)$ is reflexive and symmetric.*

Solution:

1. *R is Reflexive:* We have $\forall\ a \in I_0$, $a^a = a^a$ *i.e.*, a R a. Therefore R is reflexive.
2. *R is Symmetric:* Suppose we have a I_0 b. Then $a^b = b^a$. Now $a^b = b^c$ implies $b^a = a^b$ *i.e.*, b R a. Thus, a R b ⇒ b R a. Therefore R is symmetric.

Example 50: *Let the relation R in the set of real numbers be defined as a R b if and only if 1 + ab > 0. Show that this relation is reflexive and symmetric but not transitive.*

Solution: Let S denote the set of all real numbers. Let R be a relation in S defined as a R b iff 1 + ab > 0.

1. *R is Reflexive:* Let a be any real number.

   Then $1 + aa = 1 + a^2 > 0$, since $a^2 \geq 0$.

   Thus a R a $\forall\ a \in S$. Therefore R is reflexive.
2. *R is Symmetric:* Let a, b be any two real numbers. Then

   a R b ⇒ 1 + ab > 0

   ⇒ 1 + ba > 0. [∵ ab = ba]

   ⇒ b R a.

∴ R is symmetric.

3. *R is not Transitive:* Consider three real numbers

$$1, -\frac{1}{2}, -4$$

We have $1 + 1.\left(-\frac{1}{2}\right) = \frac{1}{2} > 0.$

$$\therefore \qquad 1\ R - \frac{1}{2}.$$

Further $1 + \left(-\frac{1}{2}\right)(-4) = 3 > 0.$

$\therefore \qquad -\frac{1}{2} \text{ R} - 4.$

But 1 + 1 (–4) = – 3 which is not greater than 0. Therefore 1 is not R-related to –4.

Thus $1 \text{ R} - \frac{1}{2}, -\frac{1}{2} \text{ R} - 4$ and 1 is not R-related to –4.

$\therefore$ R is not transitive.

Example 51: *Give an example of a relation which is:*

1. Symmetric and reflexive but not transitive.

2. Symmetric and transitive but not reflexive.

3. Reflexive but neither symmetric nor 'transitive.

Solution: Consider the set S = {1, 2, 3}. Then

1. R_1 = {(1, 1), (2, 2), (3, 3), (2, 3), (3, 2), (1, 2), (2, 1)} is a relation on S such that R_1 is reflexive and symmetric but not transitive.

   Observe that (1, 2), $\in R_1$, (2, 3) $\in R_1$ but (1, 3) $\notin R_1$ and so R_1 is not transitive.

2. R_2 = {(1, 2), (2, 2), (1, 2), (2, 1)} is a relation on S such that R_2 is symmetric and transitive but not reflexive. Observe that (3, 3) $\notin R_3$ and so R_2 is not reflexive.

3. R_3 = {(1, 1), (2, 2), (3, 3), (1, 2), (2, 3)} is a relation on S such that R_3 is reflexive but is neither symmetric nor transitive.

Example 52: *In the set N of all natural numbers, the relation R defined by aRb iff a divides b is a partial order relation.*

Solution: We have, "a $\in$ N, a is a divisor of a *i.e.,* aRa. Therefore R is reflexive.

Again, if a is a divisor of b then b cannot be a divisor of a unless a = b. Thus aRb and bRa $\Rightarrow$ a = b. Therefore R is anti-symmetric.

Finally a is a divisor of b and b is a divisor of c implies a is a divisor of c. Therefore R is transitive.

Since R is reflexive, anti-symmetric and transitive, therefore R is a partial order relation.

Example 53: *Prove that the relation of similarity in the set of all triangles in a plane is an equivalence relation.*

Solution: Let A be the set of all triangles in a plane. Let R be the relation in A defined as x R y if and only if the triangle x is similar to the triangle y, x ∈ A, y ∈ A.

R is Reflexive: Let x ∈ A. Since every triangle is similar to itself, therefore by our definition of R, we have x R x.

Thus x R x ∀ x ∈ A. Therefore R is reflexive.

R is Symmetric: Let x, y ∈ A be such that x R y. We have x R y ⇒ triangle x is similarly to triangle y [by def. of R]

⇒ triangle y is similar to triangle x

⇒ y R x [by def. of R].

Thus x r y ⇒ y R x. Therefore R is symmetric.

R is Transitive: Let x, y, z ∈ A be such that x R y and y R z. We have x R y and y R z

⇒ triangle x is similar to triangle y and triangle y is similar to triangle z

⇒ triangle x is similar to triangle z

⇒ x R z [by def. of R].

Thus, x R y and y R z ⇒ x R z. Therefore R is transitive.

Since R is reflexive, symmetric and transitive therefore R is an equivalence relation.

Example 54: *Show that the relation "greater than" denoted by > in the set of natural numbers N is transitive but is neither reflexive nor symmetric.*

Solution:

1. *R is not Reflexive:* We have 3 ∈ N, but 3 is not greater than 3 *i.e.,* 3 is not R-related to 3. Therefore R is not reflexive.
2. *R is not Symmetric:* We have 4 ∈ N, 2 ∈ N, 4 is greater than 2 *i.e.,* 4 R 2 but 2 is not greater than 4 *i.e.,* (2, 4) ∉ R. Therefore R is not symmetric.
3. *R is Transitive:* Suppose a R b and b R c. Then a > b and b > c. Now a > b and b > c implies a > c *i.e.,* a R c. Therefore R is transitive.

Example 55: *Show that corresponding to an equivalence relation R defined in a set S, for any two members be and c of the same equivalence class, b has the relation R to c.*

Solution: Suppose b and c are any two members of the equivalence class [a] = {x : x Ra, x ∈ S, a ∈ S}.

Then we have b R c and c R a.

Now c R a ⇒ a R c since R is symmetric :

Again bRa and aRc ⇒ b R c, since r is transitive. Hence the result.

Example 56: *Given that R is a relation in a set A and R^{-1} is its inverse relation, prove that*

1. *if R is reflexive, then R^{-1} is reflexive,*
2. *if R is symmetric, then R^{-1} is symmetric,*
3. *if R is transitive, then R^{-1} is transitive.*

Solution:

1. $\forall$ a $\in$ A, (a, a) $\in$ R since R is reflexive.

Now (a, a) $\in$ R ⇒ (a, a) $\in$ R^{-1}.

Thus $\forall$ a $\in$ A, (a, a) $\in$ R^{-1} and therefore R^{-1} is reflexive.

2. If (a, b) $\in$ R^{-1}, then (b, a) $\in$ R.

Since R is symmetric, therefore

$$(b, a) \in R \Rightarrow (a, b) \in R \Rightarrow (b, a) \in R^{-1}.$$

Thus (a, b) $\in$ R^{-1} ⇒ (b, a) $\in$ R^{-1} and therefore R^{-1} is symmetric.

3. Let (a, b) and (b, c) $\in$ R^{-1}. Then (c, b) $\in$ R and (b, a) $\in$ R. since R is transitive, therefore

$$(c, b) \in R \text{ and } (b, a) \in R$$

$$\Rightarrow (c, a) \in R$$

$$\Rightarrow (a, c) \in R^{-1}.$$

Thus (a, b) $\in$ R^{-1} and (b, c) $\in$ R^{-1}

⇒ (a, c) $\in$ R^{-1}.

Hence R^{-1} is transitive.

Thus, we can say that if R is an equivalence relation in A, then R^{-1} is also an equivalence relation in A.

Example 57 : *If A, B and C are any three sets, then prove that :*

(a) $A \times (B \cap C) = (A \times B) \cap (A \times C)$,

(b) $A \times (B \cup C) = (A \times B) \cup (A \times C)$,

Solution :

(a) We shall show that

$$A \times (B \cap C) \subset (A \times B) \cap (A \times C)$$

and $(A \times B) \cap (A \times C) \subset A \times (B \cap C)$

Firstly, let $(x, y) \in A \times (B \cap C)$, then

$(x, y) \in A \times (B \cap C)$

$\Rightarrow$ $x \in A$ and $y \in (B \cap C)$

$\Rightarrow$ $x \in A$ and $[y \in B$ and $y \in C]$

$\Rightarrow$ $[x \in A$ and $y \in B]$ and $[x \in A$ and $y \in C]$

$\Rightarrow$ $(x, y) \in (A \times B)$ and $(x, y) \in (A \times C)$

$\Rightarrow$ $(x, y) \in (A \times B) \cap (A \times C)$

Thus, $A \times (B \cap C) \subset (A \times B) \cap (A \times C)$ ...(1)

Again, let $(x, y) \in (A \times B) \cap (A \times C)$, then

$(x, y) \in (A \times B) \cap (A \times C)$

$\Rightarrow$ $(x, y) \in A \times B$ and $(x, y) \in (A \times C)$

$\Rightarrow$ $(x \in A, y \in B)$ and $(y \in A, y \in C)$

$\Rightarrow$ $x \in A, y \in B \cap C$

$\Rightarrow$ $(x, y) \in A \times B \cap C$

$\therefore$ $(A \times B) \cap (A \times C) \subset A (B \cap C)$ ...(2)

From (1) and (2)

$A \times (B \cap C) = (A \times B) \cap (A \times C)$

(b) Firstly, let $(x, y) \in A \times (B \cup C)$, then

$(x, y) \in A \times (B \cup C)$

$\Rightarrow$ $x \in A, y \in (B \cup C)$

$\Rightarrow$ $x \in A, [y \in B$ or $y \in C]$

$\Rightarrow$ $[x \in A, y \in B]$ or $[x \in A, y \in C]$

$\Rightarrow$ $(x, y) \in A \times B$ or $(x, y) \in A \times C$

$\Rightarrow$ $(x, y) \in (A \times B) \cup (A \times C)$.

Since, $(x, y) \in A \times (B \cup C)$

$\Rightarrow$ $(x, y) \in (A \times B) \cup (A \times C)$

$\therefore$ $A \times (B \cup C) \subset (A \times B) \cup (A \times C)$ ...(1)

Similarly, it can be proved that

$(x, y) \in (A \times B) \cup (A \times C)$

$\Rightarrow$ $(x, y) \in A \times (B \cup C)$,

$\therefore$ $(A \times B) \cup (A \times C) \subset A \times (B \cup C)$ ...(2)

From (1) and (2)

$$A \times (B \cup C) = (A \times B) \cup (A \times C)$$

Example 58 : *If $A \subset B$ show that $A \times A \subset (A \times B) \cap (B \times A)$.*

Solution : Let $(x, y) \in A \times A$, then

$(x, y) \in A \times A$,

$\Rightarrow$ $x \in A, y \in A$

Since $A \subset B$ is given,

$x \in A \Rightarrow y \in B$.

$\therefore$ $(x, y) \in A \times A$

$\Rightarrow$ $x \in A, y \in B$

$\Rightarrow$ $(x, y) \in (A \times B)$ ...(1)

Also, $(x, y) \in A \times A$

$\Rightarrow$ $x \in A, y \in A$

$\therefore$ $x \in B, y \in A$ $[\because A \subset B]$

$\Rightarrow$ $(x, y) \in B \times A$ ...(2)

Hence, from (1) and (2),

$(x, y) \in A \times A$

$\Rightarrow$ $(x, y) \in (A \times B)$ and $(x, y) \in B \times A$

$\Rightarrow$ $(x, y) \in (A \times B) \cap (B \times A)$

$\therefore$ $A \times A \subset (A \times B) \cap (B \times A)$.

Example 59 : *If $A \subset B$, prove that $A \times C \subset B \times C$ for any set C.*

Solution : Let (x, y) be any arbitrary element of $A \times C$. Then,

$(x, y) \in A \times C$

$\Rightarrow$ $x \in A$ and $y \in C$

$\Rightarrow$ $x \in B$ and $y \in C$ $[\because A \subset B]$

$\Rightarrow$ $(x, y) \in B \times C$

$\therefore$ $A \times C \subset B \times C$.

Example 60 : *Prove, using mathematical induction that $10^{2n-1} + 1$ is divisible by 11 for all $n \in N$.*

Solution : Let P(n) be the statement.

P(n) : "$10^{2n-1} + 1$ is divisible by 11."

When $n = 1$, we get $10^{2n-1} + 1 = 10 + 1 = 11$,

Which is divisible by 11. Hence, P(1) is true.

Let the result be true for P(r), *i.e.*, $10^{2r-1} + 1$ is divisible by 11. Therefore,

$$10^{2r-1} + 1 = 11k, k \in N.$$

$$\text{or} \qquad 10^{2r-1} = 11k - 1.$$

We are now to prove that P(r + 1) is true.

$$\text{Now,} \qquad = P(r + 1) : 10^{(2r+1)-1} - 1 = 10^{2r+1} + 1$$

$$= 10^{2r-1}.10^2 + 1 + (11k-1)\ 100 + 1 = 11(100k-9)$$

$$= \text{multiple of } 11.$$

Hence, 10^{2r+1} is divisible by 11. That is, P(r + 1) is true. By the principle of mathematical induction P(n) is true for all n ∈ N.

Example 61: *Which of the following relations in the set of real numbers E are equivalence relations.*

1. a, R b if | a | = | b |,
2. a R b if | a | ≥ | b |
3. a R b if a – b ≥ 0?

Solution:

1. We have, ∀ a ∈ E. | a | = | a | *i.e.,* a R a.

   Therefore R is reflexive.

   Again | a | = | b |

   ⇒ | b | = | a | *i.e.,* a R b ⇒ b R a.

   Therefore R is symmetric.

   Finally | a | = | b | and | b | = | c |

   ⇒ | a | = | c |

   i.e., a R b and b R c

   ⇒ a Rc. Therefore R is transitive.

   Hence R is an equivalence relation.
2. We have | 8 | ≥ | 7 |, but | 7 | is not ≥ | 8 |. Thus, 8 R 7 but 7 is not R-related to 8. Therefore R is not symmetric. Hence R is not an equivalence relation.
3. We have 4 – 3 ≥ 0; but 3 – 4 is not ≥ 0. Thus 4 R 3 but 3 is not R-related to 4. Therefore, R is not symmetric. Hence R is not an equivalence relation.

Example 62: *Show that if R and R' be symmetric relations in a set A, then $R \cup R'$ is also a symmetric relation in A.*

Solution: Since R and R' are relations in A, therefore $R \subseteq A \times A$ and $R' \subseteq A \times A$. Hence $R \cup R' \subseteq A \times A$ and therefore $R \cup R'$ is also a relation in A.

Let $(a, b) \in R \cup R'$.

Now $(a, b) \in R \cup R'$

$\Rightarrow$ $(a, b) \in R$ or $(a, b) \in R'$

$\Rightarrow$ $(b, a) \in R$ or $(b, a) \in R'$ [$\because$ R and R' are symmetric]

$\Rightarrow$ $(b, a) \in R \cup R'$.

Thus we have shown that

$(a, b) \in R \cup R'$

$\Rightarrow$ $(b, a) \in R \cup R'$.

Hence $R \cup R'$ is symmetric.

Example 63: *Consider the set $N \times N$, the set of ordered pairs of natural numbers. Let R be the relation in $N \times N$ which is defined by (a, b) R (c, d) if $a + d = b + c$.*

Prove that R is an equivalence relation.

Solution:

1. *R is Reflexive:* $\forall$ $(a, b) \in N \times N$, we have $a + b = b + a$ *i.e.,* (a, b) R (a, b). Therefore R is reflexive.
2. *R is Symmetric:* Suppose (a, b) R (c, d) and (c, d). Then $a + d = b+c$, which implies that $c + b = d + a$ *i.e.,* (c, d) R (a, b). Hence, R is symmetric.
3. *R is Transitive:* Now suppose (a, b) R (c, d) and (c, d) R (e, f). Then $a + d = b + c$ and $c + f = d + e$.

   Thus, $(a + d) + (c + f) = (b + c) + (d + e)$. Subtracting $c + d$ from both sides, we get $a + f = b + e$.

$\therefore$ (a, b) R (e, f) and hence r is transitive.

Since R is reflexive, symmetric and transitive, therefore R is an equivalence relation and therefore induces a partition of $N \times N$.

Example 64: *Let α be a mapping of a set S into a set T. Then if we define a R b for $a \in S$ and $b \in S$, provided $\alpha(a) = \alpha(b)$, prove that r is an equivalence relation.*

Solution: α is a mapping from S into T. If $a \in S$, then by α (a) we mean the image of a under the mapping α.

1. *R is Reflexive.* Let a be any element of S.

   Then $\alpha(a) = \alpha(a)$.

   Thus $\alpha(a) = \alpha(a)\ \forall\ a \in S$.

$\therefore$ $a\ R\ a\ \forall\ a \in S$ and so R is reflexive.

2. *R is Symmetric.* Let a, b be any two elements of S. Then a R b

$\Rightarrow$ $\alpha(a) = \alpha(b)$

$\Rightarrow$ $\alpha(b) = \alpha(a)$

$\Rightarrow$ b R a.

$\therefore$ R is symmetric.

3. *R is Transitive.* Let a, b, c be any three elements of S.

   Then $a\ R\ b \Rightarrow \alpha(a) = \alpha(b)$

   and $b\ R\ c \Rightarrow \alpha(b) = \alpha(c)$.

   Now $\alpha(a) = \alpha(b), \alpha(b) = \alpha(c)$

   $\Rightarrow$ $\alpha(a) = \alpha(c) \Rightarrow a\ R\ c$.

Thus a R b and b R c

$\Rightarrow$ a R c and so R is transitive.

Since R is reflexive, symmetric and transitive, therefore R is an equivalence relation.

Example 65: *Consider the set $N \times N$, the set of all ordered pairs of natural numbers. Let R be the relation in $N \times N$ which is defined by (a, b) R (c, d) if and only if ad = bc.*

Prove that R is an equivalence relation and therefore induces a partition of $N \times N$.

Solution:

1. *R is Reflexive:* $\forall$ (a, b) $\in N \times N$, we have ab = ba *i.e.,* (a, b) R (a, b). Therefore R is reflexive.
2. *R is Symmetric:* Suppose (a, b) R (c, d). Then ad = bc, which implies cb = da. Thus (c, d) R (a, b) and R is symmetric.
3. *R is Transitive:* Now suppose (a, b) R (c, d) and (c, d) R (e, f). Then ad = bc and cf = de.

Thus (ad) (cf) = (bc) (de).

Dividing both sides by dc, we get

af = be *i.e.,* (a, b) R (e, f).

Therefore R is transitive.

Hence R is an equivalence relation.

Example 66: *Show that the identity relation on a set S is an equivalence relation.*

Solution: Let R be the identity relation on a set S.

Then R = {(a, a) : a ∈ S}. To prove that R is an equivalence relation in S.

R is Reflexive: Let a be any element of S. Then by definition of R, we have (a, a) ∈ R ⇒ a R a.

Thus a R a ∀ a ∈ S. Therefore R is reflexive.

R is Symmetric: Let a, b, ∈ S be such that a R b. Then

(a, b) ∈ R.

Therefore we must have a = b and thus, a R b ⇒ b R a. So R is symmetric.

R is Transitive: Let a, b, c ∈ S be such that a R b and b R c. Then (a, b) ∈ R and (b, c) ∈ R. Therefore, we must have a = b and b = c. From this we get a = c and so we have (a, c) ∈ R *i.e.,* we have a R c. Thus, a R b and b R c ⇒ a R c. Therefore R is transitive.

Hence R is an equivalence relation on S.

Example 67: *If R and S are equivalence relations in a set X, prove that R ∩ S is an equivalence relation in X.*

Solution: Since R and S are relations in X, therefore R ⊆ X × X and S ⊆ X × X. Hence R ∩ S ⊆ X × X. Therefore R ∩ S is also a relation in X.

R ∩ S is reflexive. We have

(a, a) ∈ R ∀ a ∈ X. [∵ R is reflexive]

Also (a, a) ∈ S ∀ a ∈ X. [∵ S is reflexive]

∴ ∀ a ∈ X, we have

(a, b) ∈ R ∩ S. Therefore R ∩ S is reflexive.

R ∩ S is symmetric. Let (a, b) ∈ R ∩ S.

Now (a, b) ∈ R ∩ S ⇒ (a, b) ∈ R and (a, b) ∈ S

⇒ (b, a) ∈ R and (b, a) ∈ S. [∵ R and S are symmetric]

⇒ (b, a) ∈ R ∩ S.

Therefore R ∩ S is symmetric.

$R \cap S$ is transitive. We have

$(a, b) \in R \cap S$ and $(b, c) \in R \cap S$

$\Rightarrow$ $[(a, b) \in R$ and $(a, b) \in S]$

and $[(b, c) \in R$ and $(b, c) \in S]$

$\Rightarrow$ $[(a, b) \in R$ and $(b, c) \in R]$

and $[(a, b) \in S$ and $(b, c) \in S]$

$\Rightarrow$ $(a, c) \in R$ and $(a, c) \in S$ [$\because$ R and S are transitive]

$\Rightarrow$ $[(a, c) \in R \cap S$. Therefore $R \cap S$ is transitive.

Since $R \cap S$ is reflexive, symmetric and transitive, therefore $R \cap S$ is an equivalence relation.

Example 68: *Let S be the set of all point in a plane. Let R be a relation such that for any two points a and b, a R b if b is within one inch from a. Show that R is reflexive and symmetric but, not transitive.*

Solution:

1. *R is Reflexive:* Let a be any arbitrary point $\in$ S. Then distance of a from a is zero and therefore a is within one inch from a. Hence $\forall$ $a \in S$, we have a R a. Thus R is reflexive.
2. *R is Symmetric:* Suppose we have a R b *i.e.,* b is within one inch from a . Now the distance of a from b is equal to the distance of b from a. Therefore if b is within one inch from a, then a is also within one inch from b. So a R b implies b R a. Therefore R is symmetric.
3. *R is not Transitive:* Suppose, we have a R b and b R c. Then b is within one inch from a and c is within one inch from b. Suppose a, b and c are in the same straight line and b lies between a and c. Let the distance of b from a be 3/4 inch and the distance of c from b be also 3/4 inch. Then the distance of c from a is 3/4 + 3/4 *i.e.,* 1½ inches. Thus c is not within one inch from and therefore a is not R-related to c. Hence R is not transitive.

Example 69 : *Prove that a relation R on a set A is symmetric iff*

$$R = R^{-1}.$$

Solution : First, let R be a symmetric relation on set A. Then we have to prove that $R = R^{-1}$. In order to prove this we have to prove that $R \subset R^{-1}$ and $R^{-1} \subset R$.

Now, $(a, b) \in R \Rightarrow (b, a) \in R$ ($\because$ R is symmetric)

$\Rightarrow (a, b) \in R^{-1}$ (by def. of inverse relation)

Thus, $(a, b) \in R \Rightarrow (a, b) \in R^{-1}$ for all a, b $\in$ A.

So, $R \subset R^{-1}$ ...(i)

Now, let (x, y) be an arbitrary element of R^{-1}. Then

$(x, y) \in R^{-1} \Rightarrow (y, x) \in R$ (by def. of inverse relation)

$\Rightarrow (x, y) \in R$ ($\because$ R is symmetric)

Thus, $(x, y) \in R^{-1} \Rightarrow (x, y) \in R$ for all x, y $\in$ A.

So, $R^{-1} \subset R$ ...(ii)

Thus, from (i) and (ii), we get $R = R^{-1}$. *Conversely,* let R be a relation on set A such that $R = R^{-1}$. Then we have to prove that R is a symmetric relation on set A.

Let (a, b) $\in$ R. Then

$(a, b) \in R \Rightarrow (b, a) \in R^{-1}$ (by def. of inverse relation)

$\Rightarrow (b, a) \in R$, ($\because R = R^{-1}$)

Thus, $(a, b) \in R \Rightarrow (b, a) \in R$ for all a, b $\in$ A.

So, R is a symmetric relation on A.

Hence, R is symmetric iff $R = R^{-1}$.

Example 70 : *Let a relation R_1 on the set R of real numbers be defined as $(a, b) \in R_1 \Leftrightarrow 1 + ab > 0$ for all a, b $\in$ R. Show that R_1 is reflexive and symmetric but not transitive.*

Solution : (i) *Reflexivity :* Let a be an arbitrary element of R. Then

$a \in R_1 \Rightarrow 1 + a, a = 1 + a^2 > 0$ ($\because a^2 > 0$ for all a $\in$ R)

$\Rightarrow (a, a) \in R_1$ (by def. of R_1)

Thus, $(a, a) \in R_1$ for all a $\in$ R. So R_1 is reflexive on R.

(ii) *Symmetry :* Let $(a, b) \in R_1$, Then

$(a, b) \in R_1 \Rightarrow 1 + ab > 0 \Rightarrow 1 + ba > 0$

$\Rightarrow (b, a) \in R_1$ (by def. of R_1)

Thus, $(a, b) \in R_1 \Rightarrow (b, a) \in R_1$ for all a, b $\in$ R. So, R_1 is symmetric on R.

(iii) *Transitivity :* We observe that $(1, 1/2) \in R_1$ and $(1/2, 1) \in R_1$ but $(1, -1) \notin R_1$ because $1 + 1 \times (-1) = 0 \not> 0$.

So, R_1 is not transitive on R.

Example 71 : *Let N denote the set of all natural numbers and R be the relation on N × N defined by (a, b) R (c, d) ⇔ ad (b + c) = bc (a + d). Check whether R is an equivalence relation on N × N.*

Solution : (i) *Reflexivity :* Let (a, b) be an arbitrary element of N × N. Then,

$$(a, b) \in N \times N \Rightarrow a, b \in N$$
$$\Rightarrow a\, b\, (b + a) = ba\, (a + b)$$
$$\Rightarrow (a, b)\ R\ (a + b)$$

Thus, (a, b) R (a, b) for all (a, b) ∈ N × N. So, R is reflexive on N × N.

(ii) *Symmetry :* Let (a, b), (c, d) ∈ N × N be such that (a, b) R (c, d). Then

$$(a, b)\ R\ (c, d) \Rightarrow ad\, (b + c) = bc\, (a + d)$$
(by comm. of add. and mult. on N)
$$\Rightarrow (c, d)\ R\ (a, b)$$

Thus, (a, b) R (c, d) ⇒ (c, d) R (a, b) for all (a, b), (c, d) ∈ N × N. So, R is symmetric on N × N.

Transitivity : Let (a, b), (c, d), (e, f) ∈ N × N such that (a, b) R (c, d) and (c, d) R (e, f). Then

$$(a, b)\ R\ (c, d) = ad\, (b + c) = bc\, (a + d)$$

$$\Rightarrow \frac{b + c}{bc} = \frac{a + d}{ad} \Rightarrow \frac{1}{b} + \frac{1}{c} + \frac{1}{a} + \frac{1}{d} \qquad \text{...(i)}$$

and (c, d) R (e, f) ⇒ ef (d + e) = de (e + f)

$$\Rightarrow \frac{d + e}{de} = \frac{e + f}{ef} \Rightarrow \frac{1}{d} + \frac{1}{e} = \frac{1}{e} + \frac{1}{f} \qquad \text{...(ii)}$$

Adding (i) and (ii), we get

$$\left(\frac{1}{b} + \frac{1}{c}\right) + \left(\frac{1}{d} + \frac{1}{e}\right) = \left(\frac{1}{a} + \frac{1}{d}\right) + \left(\frac{1}{c} + \frac{1}{f}\right)$$

$$\Rightarrow \quad \frac{1}{b} + \frac{1}{e} = \frac{1}{a} + \frac{1}{f} \Rightarrow \frac{b+e}{be} = \frac{a + f}{af}$$

$$\Rightarrow \quad af\, (b + e) = be\, (a + f) \Rightarrow (a, b)\ R\ (e, f)$$

Thus, (a, b) R (c, d) and (c, d) R (e, f)

⇒ (a, b) R (e, f) for all (a, b), (c, d), (e, f) ∈ N × N. So, R is transitive on N × N.

Hence, R being reflexive, symmetric and transitive is an equivalence relation on N × N.

Congruence Modulo m : Let m be an arbitrary but fixed integer. Two integers a and b are said to be congruence modulo m if a – b is divisible by m and we write $a \equiv b \pmod{m}$.

Thus, $a \equiv b, \pmod{m} \Leftrightarrow a - b$ is divisible by m.

For example, $18 \equiv 3 \pmod{5}$ because 18-3 = 15 which is divisible by 5. But $25 \not\equiv 2 \pmod{4}$ because 4 is not a divisor of 25-2 = 23.

Examples 72 : *Prove that the relation 'congruence modulo m' on the set Z of all integers is an equivalence relation.*

Solution : (i) *Reflexivity :* Let a be an arbitrary integer. Then

$$a - a = 0 = 0 \times m \Rightarrow a = a \text{ is divisible by m}$$
$$\Rightarrow a \equiv a \pmod{m}$$

Thus, $a \equiv a \pmod{m}$ for all $a \in Z$. so, "congruence modulo m" is reflexive.

(ii) *Symmetry :* Let a , b, ∈ Z such that $a \equiv b \pmod{m}$

Then, $a \equiv b \pmod{m}$

$$\Rightarrow a - b \text{ is divisible by m}$$
$$\Rightarrow a - b = \lambda m \text{ for } \lambda \in Z.$$
$$\Rightarrow b - a = (-\lambda)\, m \Rightarrow b = a \text{ is divisible by m}$$
$$\Rightarrow b \equiv a \pmod{m}$$

So, "congruence modulo m" is symmetric on Z.

(iii) *Transitivity :* Let a, b, c ∈ Z such that

$a \equiv b \pmod{m}$ and $b \equiv c \pmod{m}$. Then

$a \equiv b \pmod{m} \Rightarrow a - b$ is divisible by m

$$\Rightarrow a - b = \lambda_1 m \text{ for some } \lambda_1 \in Z$$

$b \equiv c \pmod{m} \Rightarrow b - c$ is divisible by m

$$\Rightarrow b - c = \lambda_2\, m \text{ for some } \lambda_2 \in Z$$

$$\therefore \quad (a - b) + (b - c) = \lambda_1\, m + \lambda_2\, m = (\lambda_1 + \lambda_2)\, m$$
$$\Rightarrow \quad (a - c) = \lambda_3\, m, \text{ where } \lambda_3 = \lambda_1 + \lambda_2 \in Z$$
$$\Rightarrow \quad a \equiv c \pmod{m}$$

Thus, $a \equiv b \pmod{m}$ and $b \equiv c \pmod{m}$

$$\Rightarrow \quad a \equiv c \pmod{m}.$$

So, 'congruence modulo m' is transitive on Z.

Hence, 'congruence modulo m' is an equivalence relation on Z.

Examples 73: *Consider a set $A = \{p, q, r\}$ and the relation R on A defined by*

$R = \{(p, p), (p, q), (q, r), (r, r)\}$ *find :*

(a) Reflexive (R);

(b) Symmetric (R); and

(c) Transitive (R).

Solution : (a) The reflexive closure on R is obtained by adding all diagonal pairs of A × A to R which are not currently in R. Hence

reflexive (R) = $R \cup \{(q, q)\}$

$= \{(p, p), (p, q), (q, q), (q, r) (r, r)\}$

(b) The symmetric closure on R is obtained by adding all the pairs in R^{-1} to R which are not currently in R. Hence.

Symmetric (R) = $R \cup \{q, p), (r, q)\}$

$= \{(p, p), (p, q), (q, p), (q, r), (r, q) (r, r)\}$

(c) The transitive closure on R, since A has three elements, is obtained by taking the union of R with $R^2 = RoR$ and $R^3 = RoRoR$. Now,

$R^2 = RoR = \{(p, p), (p, q), (p, r), (q, r), (r, r)\}$

$R^3 = RoRoR = \{(p, p), (p, q), (p, r), (q, r), (r, r)\}$

Hence, transitive (R) = $R \cup R^2 \cup R^3$

$= \{(p, p), (p, q), (p, r), (q, r), (r, r)\}$

Examples 74 : *If $A = \{1, 2, 3, 4, 5, 6, 7\}$, which of the following two is a partition giving rise to an equivalence relation? Why?*

(i) $A_1 = \{1, 3, 5\}$, $A_2 = \{2\}$, $A_3 = \{4, 7\}$

(ii) $B_1 = \{1, 3, 5, 7\}$, $B_2 = \{3\}$, $B_3 = \{4, 6\}$

Solution : Since A is not the union of A_1, A_2 and A_3, therefore sets A_1, A_2, A_3 do not form a partition of A. Clearly, B_1, B_2 and B_3, are disjoint sets such that their union is A. So, these sets form a partition of A.

Examples 75 : *Show that in the set N of all natural numbers, the relation R defined by a R b if a divides to is a partial order relation.*

Solution : (i) We have $\forall$ a $\in$ N, a is a divisor of a, *i.e.*, a R a. Therefore, R is reflexive.

(ii) Again, if a is a divisor of b then b cannot be a divisor of a unless a = b. Thus, a R b and b R a ⇒ a = b. Therefore, R is anti-symmetric.

(iii) Finally, a is a divisor of b and b is a divisor of c implies a is a divisor of c. Therefore R is transitive.

Since R is reflexive, anti-symmetric and transitive, therefore R is a partial order relation.

MAPPING OR FUNCTION

The concept of mapping of one set into another is of great importance in mathematics. It is not a new concept to any of us as we have been considering mapping from the beginning of our mathematical training. For example, plotting of the relation $y = x^2$ is nothing but to study the particular mapping which takes every real number into its square. The following discussion will make the concept of mapping clear.

Suppose A and B are any two non-empty sets. Let A = {a, b, c, d}, B = {x, y, z}. Suppose by some rule or other, we assign to each element of A unique element of B. Suppose a is associated to x, b is associated to y, c is associated to x and d is associated to z. The set of such assignments is called a *'function'* or *'mapping'* from A to B. If we denote this set by f than we write

$$f : A \to B$$

Which is read as *"f is a function of A to B"* or *"f is a mapping from A to B"*.

Loosely speaking, a mapping from one set into another may be defined as given below :

Definition : *Let A and B be two given sets. Suppose there exists a rule denoted by f, which associates to each member of A, a unique member of B. Then f is called a function or a mapping of A to B. The mapping f of A to B is denoted by* $f : A \to B$ *or by* $A \xrightarrow{f} B$.

Further, if a ∈ A, then the element in B which is assigned to a is called the f image of a or the value of the function f for a and is denoted by f(a). Now, we think of a definition which serves to make the concept of a mapping precise and thus formally, we define mapping as follows:

Definition : *If A and B are nom-empty sets then a mapping from A to B is a subset C of A × B such that for every a ∈ A there is a unique b ∈ B such that the ordered pair (a, b) is in C.* For all practical purposes the following notion of mapping will be used hereafter and thus we arrive at another form of definition.

Definition : *A mapping $f : A \rightarrow B$, is a rule which associates any element $a \in A$ with some element $b \in B$, the rule being associate (or map) $a \in A$ with $b \in B$ if $(a, b) \in C$, C being the subset of $A \times B$.*

Note : The rule f should possess the characteristic that there may be some elements of the set B which are not associated to any element of the set A but each element of the set A must be associated to one and only one element of the set B. Two or more elements of the set A may be associated to the same element of the set B but association of one element of A to more than one element in B is not permissible.

Domain, Co-domain and Range of Functions : Let f be a mapping of A into B. Then A is called the *'domain'* of the function f and B the *'co-domain'* of the function f. It is evident from the definition that each element of B need not appear as the image of an element in A. We define the *'range'* of f to consist of all those elements in B which appear as f image of atleast one element in A. There can be more than one elements of A which have the same image in B. The image set f [A] is called the range of f.

Functions Defined as Sets of Ordered Pairs : Let A and B are any two non-empty sets, then a mapping f of A to B is a subset f of $A \times B$ satisfying the following conditions :

(i) for each $a \in A$, $(a, b) \in f$, for some $b \in B$;

(ii) if $(a, b) \in f$ and $(a, b') \in f$, then $b = b'$.

The first condition ensures that we have a rule that assigns to each element a ε A some element b ε B. Thus each element in A will have image The second condition guarantees that the image is unique. Accordingly, f is a function from A to B.

Note : If $f : A \rightarrow B$, it is important to distinguish between a function f and the value f(x) of f for any element x. While f is a subset of $A \times B$, f(x) is an element of the set B.

Operator : *If the domain and co-domain of a function of are both the same set, say*

$$f : A \rightarrow A$$

the f is called an operator or transformation on A.

Following examples will make the notion of a function more clear;

Examples :

(i) Let A = {1, 2, 3, 4}

and B = {1, 2, 3, 4, 5, 6, 7, 8, 9, 10, 11, 12}

Now, let f assign to each number in A is square in B. Then f is not a mapping from A to B since no number of B is assigned to the element $4 \in A$.

(ii) Let X = {p, q, r}, Y = {1, 2, 3} and mapping f is as follows :

$f(p) = 3, f(q) = 1, f(r) = 2,$

then f-image of X is {3, 1, 2} or in other words f-image of X is Y.

(iii) Let f be a mapping of N into N such that

$f(1) = 3; f(2) = 5; f(3) = 7; ...$

We can express this mapping by the functional notion as

$f : N \to N$; defined by $f(x) = 2x + 1, \forall x \in N$.

(iv) Let A be the set of countries in the world and B be the set of capital cities then (i) every country has a capital assigned to it, (ii) no country will have two capitals. Thus, f is a function from the set A to B and the image of India under f is New Delhi, *i.e.*, f(India) = New Delhi. Similarly, f(Nepal) = Kathmandu.

Here, domain of f is the list of countries in the world and the co-domain is the list of capital cities of these countries.

Diagrammatic Representation of Function : Sometimes a function may be represented by a diagram as will be obvious from the following example :

Let $A = \{a, b, c, d\}$ and

$B = \{u, x, y, z\}$.

Let $f : A \to B$ defined by the correspondence $f(a) = y$, $f(b) = x$, $f(c) = z$ and $f(d) = y$.

We represent the two sets A and B by the points inside the two circles. The mapping $f : A \to B$ is represented by means of arrows joining the points which represent the elements of A to points representing the corresponding elements of B. By the definition of function, it is obvious that.

(i) every point in A is joined to some point in B by an arrow.

(ii) a point in A cannot be joined to two or more distinct points in B.

(iii) two or more points in A may be joined to the same point in B.

(iv) there may be some points in B which are not joined to any point in A.

Equal Functions : *Two functions f and g are said to be equal iff :*

(i) the domain of f = domain of g

(ii) the co-domain of f = the co-domain of g and

(iii) f(x) = g(x) for every x belonging to their common domain.

Example : Let A = {1, 2}, B = {3, 6}, and f : A → B given by f(x) = $x^2 + 2$ and g : A → B given by g(x) = 3x. Then, we observe that f and g have the same domain and co-domain and also

$$f(1) = 3 = g(1) \text{ and } f(2) = 6 = g(2).$$

Hence f = g.

FUNCTION AS A RELATION

Let A and B be two non-empty sets. A relation f from A to B, i.e., a sub-set of A × B, is called a function (or a mapping or a map) from A to B if :

(i) for each a ∈ A there exists b ∈ B such that (a, b) ∈ f,

(ii) (a, b) ∈ f and (a, c) ∈ f ⇒ b = c.

Thus, a non-void subset f of A × B is a function from A to B if each element of A appears in some ordered pair in f and no two ordered pairs in f have the same first element.

If (a, b) ∈ f, then b is called the image of a under f.

Example : Let A = {1, 2, 3},

B = {2, 3, 4}

and f, g and h we three sub-sets of A × B as given below :

f = {(1, 2), (2, 3), (3, 4)},

g = {(1, 2), (1, 3), (2, 3), (3, 4)},

h = {(1, 3), (2, 4)}.

Then f is a function from A to B but g and h are not functions from A to B, because 1 ∈ A has two images 2 and 3 in B and h is not a function from A to B because 3 ∈ A has no image in B.

KINDS OF FUNCTIONS

If f : A → B is a function, then f associates all elements of set A to elements in set B such that an element of set A is associated to a unique element of set B. Following these two conditions we may associate :

(i) different elements of set A to different elements of set B, or

(ii) more than one element of set A may be associated to the same element of set B, or

(iii) there may be some elements is B which do not have their pre-images in A or

(iv) all elements in B may have their pre-images in A.

Corresponding to each of these possibilities we define a type of a function as given below :

One-one Function (Injection) : *A function $f : A \to B$ is said to be a one-one function or an injection if different elements of A have different images in B.*

Thus, $f : A \to B$ is one-one $\Leftrightarrow a \neq b$

$\Rightarrow$ $f(a) \neq f(b)$ for all a, b, $\in$ A

$\Leftrightarrow$ $f(a) = f(b)$

$\Rightarrow a = b$ for all a, b $\in$ A.

Examples :

(i) A function which associates to each country in the world, its capital, is one-one because different countries have their different capitals.

(ii) Let X = {1, 2, 3, 4}, Y = {1, 4, 9, 16} and $f : X \to Y$ s.t. $f(x) = x^2 \; \forall \; x \in X$, then f is one-one mapping of X into Y, as no two distinct elements of X have the same f-image in Y.

Let $f : A \to B$ be a function such that A is an infinite set and we wish to check the injectivity of f. In such a case it is not possible to list the images of all elements of set A to see whether different elements of A have different images or not. The following method provides a systematic procedure to check the injectivity of a function.

Method to Check the Injectivity of a Function :

(i) Take two arbitrary elements a, b (say) in the domain of f.

(ii) Put $f(a) = f(b)$

(iii) Solve $f(a) = f(b)$. If $f(a) = f(b)$ gives $a = b$ only, then $f : A \to B$ is a one-one function (or an injection) otherwise not.

Note : Let $f : A \to B$ and let a, b $\in$ A. Then, $a = b \Rightarrow f(a) = f(b)$ is always true from the definition. But, $f(a) = f(b) \Rightarrow a = b$ is true only when f is injective.

Many-one Function : A function $f : A \to B$ is said to be a many-one function of two or more elements of set A have the same image is B.

Thus, $f : A \to B$ is a many one function if there exist x, y $\in$ A such that $x \neq y$ but $f(x) = f(y)$.

In other words, $f : A \to B$ is a many-one function if it is not a one-one function :

Examples :

(i) $A = \{-1, 1, -2, 2\}$ and $B = \{1, 4, 9, 16\}$. Consider $f : A \to B$ s.t. $f(x) = x^2$. Then $f(-1) = 1$, $f(1) = 1$, $f(-2) = 4$, $f(2) = 4$.

Clearly 1 and – 1 have the same image. Similarly, 2 and – 2 also have the same image. So, f is a many-one function.

(ii) Consider a function $f : Z \to Z$ given by $f(x) = |x|$.

Then f is a many-one function because for every $a \in Z$, $a \neq 0$, $a \neq -a$ but $f(a) = f(-a)$ $[\because |a| = |-a|]$.

Onto Function (Surjection) : *A function $f : A \to B$ is said to be an onto function or a surjection if every element of B is the f-image of some element of A, i.e., if $f(A) = B$ or range of f is the co-domain of f.*

Thus, $f : A \to B$ is a surjection iff for each $b \in B$, $\exists\, a \in A$ such that $f(a) = b$.

Into Function : *A function of $f : A \to B$ is an into function if there exists an element in B having no pre-image in A.*

In other words, $f : A \to B$ is into function if it is not an onto function.

Examples :

(i) Let $A = \{-1, 1, 2, -2\}$, $B = \{1, 4\}$ and $f : A \to B$ be a function defined by $f(x) = x^2$. The f is onto because $f(A) = \{f(-1), f(1), f(2), f(-2)\} = \{1, 4\} = B$.

(ii) A function $f : N \to N$ defined by $f(x) = 2x$ is not an onto function, because $f(N) = \{2, 4, 6, ...\} \neq N$ (co-domain).

Method for Checking the Surjectivity of a Function : Let $f : A \to B$ be the given function.

(i) Choose an arbitrary element b in B.

(ii) Put $f(a) = b$.

(iii) Solve the equation $f(a) = b$ for a and obtain a in terms of b. Let $a = g(b)$.

(iv) If for all values of $b \in B$, the values of a obtained from $a = g(b)$ are in A, then f is onto.

If there are some $b \in B$ for which a given by $a = g(b)$, is not in A. Then f is not onto.

Bijection (One-one Onto Function) : *A function $f : A \to B$ is a bijection if it is one-one as well as onto.*

In other words, a function $f : A \to B$ is a bijection if :

(i) it is one-one, i.e., $f(x) = f(y) \Rightarrow x = y$ for all $x, y \in A$.

(ii) it is onto, i.e., for all $y \in B$, there exists $x \in A$ such that $f(x) = y$.

Examples : Let A set of even integers and B be the set of odd integers, then the mapping $f : A \to B$ given by

$$f(x) = x + 1, \forall x \in A$$

is one-one onto.

One-one into Mapping : *Any mapping which is one-one as well as into is called one-one into mapping.*

Example : Let X be the set of integers and Y the set of all even integers, then the mapping $f : X \to Y$, s.t. $f(x) = 2x$, $x \in X$ is an into mapping which is also one-one.

Many-one into Mapping : *A function which is many one as well as into is called many-one into mapping.*

Example : If $X = (x_1, x_2, x_3)$ and $Y = (y_1, y_2)$ and if the function $f : X \to Y$, is defined as $f(x_1) = y_1$, $f(x_2) = y_1$, $f(x_3) = y_1$, then it is a many-one into function.

Many-one Onto Mapping : *A function which is many one as well as onto is called many-one onto mapping.*

Example : Let $X = \{a, b, c, d\}$, $Y = \{3, 4, 5\}$ and $f(a) = 3$, $f(b) = 4$, $f(c) = 4$, $f(d) = 5$, then it is many-one onto mapping.

DIFFERENT TYPES OF MAPPINGS

Explanation from a Daily Life Example : Let A be the set of students setting on chairs in a class room and let B be the set of chairs in the classroom. Let f be the correspondence which associates to each student the chair on which he sits.

Since every student has some chair to sit on (of course two or more than two students might sit on one chair) and no student can sit on two or more than two chairs, therefore $f : A \to B$.

The following cases arise :

(i) If every student gets a separate chair and no chair is left vacant, then this is a case of *one-one onto mapping.*

(ii) If every student gets a separate chair and still some chairs lie vacant it is a *one-one into mapping.*

(iii) If every student does not sit on separate chair and no chair is left vacant then this is case of *many-one onto mapping.*

(iv) If every student does not sit on a separate chair and some chair are left vacant, then this is a case of *many-one into mapping.*

IDENTITY MAP OR IDENTITY FUNCTION

Let X be any set and the function $f : X \to X$ be defined by the formula $f(x) = x\ \forall x \in X$, i.e., each element of X is mapped on itself, then f is called the identity map or the identity function (transformation) on X.

We denote this function by I_x usually. Thus if I_x denotes the identity mapping on a set X, we have

$$I_x(x) = x,\ \forall\ x \in X.$$

Example : Let A = {a, b, c, d}. Then f = {(a, a), (b, b), (c, c), (d, d)} is an identity mapping of A.

Identity mapping is always one-one onto.

CARDINALLY EQUIVALENT SETS

Let there be a mapping $f : X \to Y$ which is one-one and onto then the two sets X and Y are said to be *cardinally equivalent* or *equinumerous.* The fact is denoted by writing $X \sim Y$.

Note : If any X is equivalent to N, the set of natural numbers then X is said to be *denumerable set.*

(i) Let X = {1, 2, 3}, Y = {1, 4, 9} then $f : X \to Y$ defined by $f(x) = x^2\ \forall\ x \in X$ is one-one and onto mapping, therefore the set X and Y are cardinally equivalent, *i.e.*, $X \sim Y$.

(ii) The set $A = \left\{1, \frac{1}{2}, \frac{1}{3}, ..., \frac{1}{n}, ...\right\}$ is denumerable set as it can be put in one-to-one correspondence with the set of natural numbers.

INCLUSION MAP

If $X \subset Y$, the function $f : X \to Y$, defined by $f(x) = x$ for each $x \in X$, is called the inclusion map.

Example : Let X = {– 3, – 2, – 1, 0, 1, 2, 3} and Y = {.... – 4, – 3, – 2, – 1, 0, 1, 2, 3, 4, ...} and f(– 3) = – 3, f(– 2) = – 2, f(– 1) = – 1, f(0) = 0, f(1) = 1, f(2) = 2, f(3) = 3, then it is an inclusion map of X to Y because $X \subset Y$ and $f(x) = x\ \forall\ x \in X$.

INCLUSION MAP

If $X \subset Y$, the function $f : X \to Y$, defined by $f(x) = x$ for each $x \in X$, is called the inclusion map.

Example : Let X = {– 3, – 2, – 1, 0, 1, 2, 3} and Y = {.... – 4, – 3, – 2, – 1, 0, 1, 2, 3, 4, ...} and f(– 3) = – 3, f(– 2) = – 2, f(– 1) = – 1, f(0) = 0, f(1) = 1, f(2) = 2, f(3) = 3, then it is an inclusion map of X to Y because $X \subset Y$ and $f(x) = x \ \forall \ x \in X$.

INVERSE IMAGE OF AN ELEMENT

Let f be a function defined from the set X to the set Y then the inverse image of an element $b \in Y$ under f is denoted by $f^{-1}(b)$ to be read as f-image b and

$$f^{-1}(b) = \{x : x \in X \text{ and } f(x) = b\}$$

i.e., $f^{-1}(b)$ is the set of those elements in X which have b as their f-image.

Example : Let $f : R \rightarrow R$ where $f(x) = x^2$ and R is the set of real numbers, then $f^{-1}(9) = \{3, -3\}$, for 9 is the f-image of both 3 and – 3.

Inverse Image of a Subset : Let f be a function defined from the set X to the set Y and B be a subset of Y, *i.e.*, $B \subset Y$ then the inverse of B under f is given by $f^{-1}(B) = \{x : x \in X \text{ and } f(x) \in B\}$.

Obviously,

$$f^{-1}(B) \subset X \text{ and } f^{-1}[f(x)] = X.$$

Note : If $b \subset Y$, then $\{b\} \subset Y$ and $f^{-1}(b) = f^{-1}\{(b)\}$.

Example : Let $f : X \rightarrow Y$, where X = {1, 2, 3, 4 5} and Y = {1, 2, 4, 6, 9, 15. 16, 25, 30} and let $f(x) = x^2 \ \forall \ x \in X$, let B_1 = {2, 4, 9, 15, 16} and B_2 = {2, 6}. Then $f^{-1}(B_1) = \{2, 3, 4\}$ and $f^{-1}(B_2) = \phi$.

INVERSE MAPPING (OR FUNCTION)

Let X and Y be two sets and let $f : X \rightarrow Y$ be a function. If we follow a rule in which elements of Y are associated to their pre-images, then we find that under such a rule there may be some elements in Y which are not associated to elements in A.

This happens when f is not an onto map. Therefore, all elements in Y will be associated to some elements in X if f is an *onto map.* Also, if it is a many-one function then under the said rule an element in Y may be associated to more than are element in X. Therefore an element in Y will be associated to a unique element in X if f is a *injective map.*

It follows from the above discussion that is $f : X \rightarrow Y$ is a bijection, we can define a new function from Y to X which associates each element $y \in Y$ to its pre-image $f^{-1}(y) \in X$. Such a function is known as the inverse of function f and is denoted by f^{-1}.

Definition : *Let $f : X \to Y$ be a bijection. Then a function $g : Y \to X$ which associates each element $y \in Y$ to a unique element $x \in X$ such that $f(x) = y$ is called the inverse of f, i.e.,*

$$f(x) = y \Leftrightarrow g(y) = x.$$

The inverse of f is generally denoted by f^{-1}.

Thus, if $f : X \to Y$ is a bijection, than $f^{-1} : Y \to X$ is such that

$$f(x) = y \Leftrightarrow f^{-1}(y) = x.$$

Example : If A = {1, 2, 3, 4,}, B = {2, 4, 6, 8} and $f : A \to B$ is given by f(x) = 2x, then

$$f(1) = 2,\ f(2) = 4,\ f(3) = 6 \text{ and } f(4) = 8$$

So, $\quad f = \{(1, 2), (2, 4), (3, 6), (4, 8)\}$

Which is clearly a bijection.

$\therefore \quad f^{-1} = \{(2, 1), (4, 2), (6, 3), (8, 4)\}$

Method to Find the Inverse of a Bijection : Let $f: X \to Y$ be a bijection. To find the inverse of f we proceed as follows :

(i) Put f(x) = y where $y \in Y$ and $x \in X$.

(ii) Solve f(x) = y to obtain x in terms of y.

(iii) In the relation obtained in step (ii) replace x by $f^{-1}(y)$ to obtain the inverse of f.

SOME RESULTS ON INVERSIBLE FUNCTIONS

Theorem 1 : *Let f be a one-one mapping of X onto Y. Then f^{-1} is uniquely defined from Y to X, i.e., the inverse of an invertible mapping is unique.*

Proof : Let g and h be two inverse mappings of f. We must then prove that g = h, for which we must prove that $g(y) = h(y)\ \forall\ y \in Y$. Let y be an arbitrary element of Y and let $g(y) = x_1$ and $h(y) = x_2$. Since g is an inverse mapping of f we have, $g(y) = x_1 \Leftrightarrow f(x_1) = y$.

Similarly h being an inverse of f, $h(y) = x_2 \Leftrightarrow f(x_2) = y$.

$\therefore \quad f(x_1) = f(x_2)$ [Each being equal to y]

But $f(x_1) = f(x_2) \Rightarrow x_1 = x_2$ [$\because$ f is one-one]

$\therefore\ g(y) = h(y)\ \forall\ y \in Y \qquad [\because x_1 = g(y) \text{ and } x_2 = h(y)]$

Hence g = h, *i.e.*, inverse of f is unique.

Theorem 2 : *Let X and Y be two non-empty sets. Let f be a one-one mapping of X onto Y. Then f^{-1} is also one-one onto i.e., inverse of an invertible mapping is invertible.*

Proof : Let f be an invertible mapping, so that f is one-one onto. Let y_1, y_2 be any two elements in Y. Then f being one-one onto, there exist unique elements x_1 and x_2 in X such that

$$f(x_1) = y_1 \text{ and } f(x_2) = y_2$$

and therefore $\quad x_1 = f^{-1}(y_1)$ and $x_2 = f^{-1}(y_2)$.

Now, $\quad f^{-1}(y_1) = f^{-1}(y_2) \Rightarrow x_1 = x_2$

$\Rightarrow \quad f(x_1) = f(x_2) \Rightarrow y_1 = y_2$

This shows that f^{-1} is one-one.

Now, let x be an arbitrary element of X. Then by definition of f there exists a unique element $y \in Y$ such that $f(x) = y$ or $x = f^{-1}(y)$. Thus for each $x \in X$, there exists a unique element $y \in Y$ such that $f^{-1}(y) = x$. So f^{-1} is onto.

Hence, f^{-1} is one-one and onto and therefore invertible.

PRODUCT OF MAPPINGS OR COMPOSITE OF FUNCTIONS

The idea of a function of a function in mathematics has an analogue in abstract mathematics and it is known as a composite function. Let X, Y, Z be three sets and f be a function defined from X to Y and g be a function defined from Y to Z.

i.e., $f : X \rightarrow Y$ and $g : Y \rightarrow Z$.

By $f : X \rightarrow Y$ we mean that to every element $x \in X$, there corresponds a unique element $f(x) \in Y$. Since the domain of g is Y, so by the function $g : Y \rightarrow Z$ we mean that to every element $f(x) \in Y$ there corresponds a unique element of $g[f(x)] \in Z$. Thus we notice that to every element $x \in X$ there corresponds a unique element $g[f(x)] \in Z$ under the mappings f and g. This implies that there exists a mapping from X to Z. This mapping is called the composite mapping or the product mapping of f and g and is denoted by *g of* or gf.

Definition : *Let $f : X \rightarrow Y$ and $g : Y \rightarrow Z$; then the composite of the functions f and g denoted by g of or gf is mapping $g\,of : X \rightarrow Z$ s.t. $(g\,of)(x) = g[f(x)]$, $\forall x \in X$.*

Example : Let $f : R \rightarrow R$ be given by the relation $f(x) = \sin x \ \forall x \in R$ and the mapping $g : R \rightarrow R$ be given by the relation

$$g(x) = x^2, \ \forall x \in R,$$

then the composite function $(g\,of) : R \rightarrow R$ is given by the relation.

$$(g\,of)(x) = g[f(x)] = g[\sin x] = (\sin x)^2 = \sin^2 x.$$

Also, $(fog)(x) = f[g(x)] = f[x^2]$

$= \sin x^2 \ \forall \ x \in R.$

Notes :

(i) If $f : X \to X$ and $g : X \to X$, then both the composite of functions g of and fog are defined. But in general gof $\neq$ fog.

(ii) It may be noted that the product (gof) is defined only when range (f) $\subset$ dom (g). When gof is defined then it is not necessary that (fog) is also defined. However, (fog) is defined when, range (g) $\subset$ dom (f).

SOME RESULTS ON COMPOSITE MAPPINGS

Theorem 1:

The product of any invertible mapping f with its inverse mapping f^{1} is an identity mapping.

Proof :

Let f be a one-one mapping of X and Y and let I_x and I_y be the identity mapping an X and Y respectively. Clearly, f^{-1} is a one-one mapping of Y onto X.

Now, $\quad f : X \to Y$

and $\quad f^{-1} : Y \to X,$

then $\quad f^{-1} \circ f : X \to X$

Now, let x be an arbitrary element of X and let $f(x) = y$

so that, $x = f^{-1}(y)$.

$$\therefore \quad (f^{-1} \circ f)(x) = f^{-1}[f(x)] = f^{-1}(y)$$
$$= x = I_x(x).$$

So $f^{-1} \circ f = I_x$.

Again, $f^{-1} : Y \to X$

and $\quad f : X \to Y,$

so $\quad f \circ f^{-1} = Y \to Y.$

Now, for an arbitrary $y \in Y$, there is associated a unique $x \in X$ such that $f(x) = y$ or $x = f^{-1}(y)$.

$$\therefore \quad (f \circ f^{-1})(y) = f[f^{-1}(y)] = f(x)$$
$$= y = I_y(y)$$

Thus $f \circ f^{-1} = I_y$.

Theorem 2:

If $f : X \to Y$ be a one-one and onto mapping, then

$$f \circ f^{-1} = I_Y$$

and $$f^{-1} \circ f = I_X,$$

where I_X and I_Y are the identity mappings of the sets X and Y respectively.

Proof:

We shall first prove that $f^{-1} f = I_X$.

Since $f : X \rightarrow y$ and $f^{-1} : Y \rightarrow X$, therefore $(f^{-1} \circ f) : X \rightarrow X$.

Let x be any arbitrary element $\in$ X and let f (x) = y, where y $\in$ Y. Then by the definition of the mapping f^{-1}, $f^{-1}(y) = x$.

We have $(f^{-1} \circ f)(x) = f^{-1}[f(x)] = f^{-1}(y) = x$. Thus the function $f^{-1} \circ f$ maps very element x $\in$ X onto itself. Therefore $f^{-1} \circ f = I_X$.

Now we shall prove that $f \circ f^{-1} = I_Y$. Since $f^{-1} : Y \rightarrow X$ and $f : X \rightarrow Y$, therefore $((f \circ f^{-1}), Y \rightarrow Y$.

Let y be any arbitrary element $\in$ Y and let $f^{-1}(y) = x$, where x $\in$ X Then f (x) = y.

We have $(f \circ f^{-1})(y) = f[f^{-1}(y)] = f(x) = y$. Thus, the function fof^{-1} maps every element y $\in$ Y onto itself. Therefore $f \circ f^{-1} = I_Y$.

Theorem 3:

Let X and Y be two non-empty sets. Let f be a one-one mapping of X onto Y. Then f^{-1} is also one-one onto i.e., inverse of an invertible mapping is invertible.

Proof :

Let f be an invertible mapping, so that f is one-one onto. Let y_1, y_2 be any two elements in Y. Then f being one-one onto, there exist unique elements x_1 and x_2 in X such that

$$f(x_1) = y_1 \text{ and } f(x_2) = y_2$$

and therefore $$x_1 = f^{-1}(y_1) \text{ and } x_2 = f^{-1}(y_2).$$

Now, $$f^{-1}(y_1) = f^{-1}(y_2) \Rightarrow x_1 = x_2$$

$\Rightarrow$ $$f(x_1) = f(x_2) \Rightarrow y_1 = y_2$$

This shows that f^{-1} is one-one.

Now, let x be an arbitrary element of X. Then by definition of f there exists a unique element y $\in$ Y such that f(x) = y or $x = f^{-1}(y)$. Thus, for each x $\in$ X, there exists a unique element y $\in$ Y such that $f^{-1}(y) = x$. So f^{-1} is onto.

Hence, f^{-1} is one-one and onto and therefore invertible.

Theorem 4:

Let $f : X \rightarrow Y$ and $g : Y \rightarrow X$. Then g is the inverse function of f i.e., $g f^{1}$, if the product function $(g \circ f) : Y \rightarrow X$ is the identity function on Y.

Proof:

In order to show that f is invertible, we are to show that f is one-one and onto.

1. *f is one-one:* Let $x_1, x_2 \in X$. Then

$$f(x_1) = f(x_2) \Rightarrow g[f(x_1)] = g[f(x_2)]$$
$$\Rightarrow (g \circ f)(x_1) = (g \circ f)(x_2)$$
$$\Rightarrow I(x_1) = 1(x_2) \qquad [\because g \circ f = I]$$
$$\Rightarrow x_1 = x_2$$

$\therefore$ f is one-one

2. *f is onto:* Let y be any element of Y. Since g is a function from Y to X, therefore $g(y) \in X$

Let $g(y) = x$. Then

$$g(y) = x \Rightarrow f[g(y)] = f(x)$$
$$\Rightarrow (f \circ g)(y) = f(x)$$
$$\Rightarrow I(y) = f(x) \qquad [\because g \circ f = I]$$
$$\Rightarrow y = f(x).$$

Thus $y \in Y \Rightarrow \# x \in X$ such that $f(x) = y$. Therefore f is onto. Since f is 1-1 onto, therefore f is invertible *i.e.,* f^{-1} exists.

3. Now we shall prove that $f^{-1} = g$. We have

$$f \circ g = I \Rightarrow f^{-1} \circ (f \circ g) = f^{-1} \circ I$$
$$\Rightarrow (f^{-1} \circ f) \circ g = f^{-1} \Rightarrow I \circ g = f^{-1} \Rightarrow g = f^{-1}.$$

Theorem 5:

Associativity of Composites of Functions

Let $f : A \rightarrow B$, $g : B \rightarrow C$ and $h : C \rightarrow D$, then

$$(h \circ g) \circ f = h \circ (g \circ f).$$

Proof:

It can be easily seen that both $(h \circ g) \circ f$ and $h \circ (g \circ f)$ are mappings of A to C. These two mappings will be equal if they assign the same image to each element x in the domain A, *i.e.,* if

$$[(h \circ g) \circ f](x) = [h \circ (g \circ f)](x).$$

We have, be definition of composites of mappings,

$$[(h \circ g) f] (x) = (h \circ g) [f (x)] = h [g \{f (x)\}] = h [(g \circ f) (x)]$$
$$= [h \circ (g \circ f)] (x).$$

Hence $(h \circ g) \circ f = h \circ (g \circ f)$.

Theorem 6:

If $f : X \to Y$, and $g : Y \to Z$ be two one-one onto maps, then $g \circ f : X \to Z$ is also one-one onto and

$$(g \circ f)^{-1} : Z \to X = (f^{-1} \circ g^{-1}) : Z \to X.$$

Proof:

We shall prove that

(i) $g \circ f$ is one-one, (ii) $g \circ f$ is onto, (iii) $(g \circ f)^{-1} \circ g^{-1}$.

(ii) Let x_1 and x_2 be any two elements in X. The mapping $g \circ f$ will be one-one if we show that the images of x_1 and x_2 under this mapping are equal only if $x_1 = x_2$. We have

$(g \circ f) (x_1) = (g \circ f) (x_2) \Rightarrow g [f (x_1)] = g [(x_2)]$

[by the definition of the composite mapping $g \circ f$]

$\Rightarrow f (x_1) = f (x_2)$ [$\because$ $f (x_1), f (x_2), \in Y$ and the mapping $g : Y \to Z$ is one-one]

$\Rightarrow x_1 = x_2$ [$\because$ f is one-one]

$\therefore$ the mapping $g \circ f$ is one-one.

(ii) Let z be any arbitrary element of Z. Since g is an onto mapping of Y to Z, therefore $\exists\ y \in Y$, such that $\exists\ x \in X$, such the $f (x) = y$. Thus, for every $z \in Z\ \exists\ x \in X$, such that $z = g (y) = g [f (x) = (g \circ f) (x)$. Therefore $g \circ f$ is a mapping of X onto Z.

Thus $g \circ f$ is one-one onto and hence is inversible.

(iii) Let z be any arbitrary element $\in Z$,

such that $z = g (y)$, $(y \in Y, z \in Z)$.

Then $y = g^{-1} (z)$.

Also let $y = f (x)$, $(y \in Y, x \in X)$. Then $x = f^{-1} (y)$.

We have $(g \circ f) (x) = g [f (x)] = g (y) = z$.

Since $g \circ f$ if one-one onto.

$\therefore$ $(g \circ f) (x) = z \Rightarrow (g \circ f)^{-1} (z) = x$.

Also $(f^{-1} \circ g^{-1}) (z) = f^{-1} [g^{-1} (z)] = f^{-1} (y) = x$.

Thus " $z \in Z, (g \circ f)^{-1}(z) = (f^{-1} \circ g^{-1})(z)$.

Hence the mappings $(g \circ f)^{-1}$ and $f^{-1} \circ g^{-1}$ are equal.

Theorem 7:

If $f : X \to Y$, then $I_Y \circ f = f$ and $f \circ I_X = f$ i.e., then product of any function with the identity function is the function itself.

Proof:

Let x be any arbitrary element $\in$ X and let

$$f(x) = y, (x \in X, y \in Y).$$

Since $f : X \to Y$ and $I_Y : Y \to X$, therefore $(I_Y \circ f) : X \to Y$.

We have $(I_Y \circ f)(x) = I_Y[f(x)] = I_Y(y) = y$.

[$\because$ I_Y is the identity mapping of Y]

Also $f(x) = y$.

Thus $\forall x \in X, (I_Y \circ f)(x) = f(x)$. Therefore $I_Y \circ f = f$.

Again since $I_x : X \to X$ and $f : X \to Y$, therefore $f \circ I_X : X \to Y$.

We have $(f \circ I_X)(x) = f[I_X(x)] = f(x)$

[$\because$ I_X is the identity mapping of X]

Thus $\forall x \in X, (f \circ I_X)(x) = f(x)$. Therefore $f \circ I_X = f$.

Theorem 8:

Composite of functions is associative.

Proof :

Let X, Y, Z, S be the four non-empty sets.

Let $f : X \to Y$, $g : Y \to Z$ and $h : Z \to S$. Then we have to show that, (h o g) of = h o (g o f).

Let x be an arbitrary element of x. Then

$$[(hog) of](x) = (hog)[f(x)]$$
$$= h[g(f(x)] = h[(gof)(x)]$$
$$= [ho(gof)](x)$$

Hence, (hog) of = ho (gof).

Theorem 9:

Let X, Y and Z be any non-empty sets and let f and g be one-one mappings of X onto Y and Y onto Z respectively so that f and g are both invertible.

Then (gof) is also invertible and

$$(gof)^{-1} = f^{-1}og^{-1}.$$

Proof :

In order to show that (gof) is invertible, we must show that, it is one-one and onto.

Now, $g(gof)(x_1) = (gof)(x_2)$

$\Rightarrow$ $f[f(x_1)] = g[f(x_2)]$

$\Rightarrow$ $g(x_1) = f(x_2)$ [$\because$ g is one-one]

$\Rightarrow$ $x_1 = x_2$ [$\because$ f is one-one]

Hence (gof) is one-one.

Now, in order to show that (gof is onto, let z be an arbitrary element of Z. Then g being onto $\exists$ an element of in Y such that g(y) = z. Also, f being onto, corresponding to the element $y \in Y$ $\exists$ an element x in X,

such that, $f(x) - y$.

$\therefore$ $(gof)(x) = g[f(x)] = g(y) = z$.

Thus, corresponding to the element $z \in Z$, $\exists$ an element $x \in X$ such that (gof) (x) = z. Consequently, (gof) is onto.

Thus, (gof) is one-one onto and hence invertible.

Now $(gof)(x) = z \Rightarrow (gof)^{-1}(z) = x$ ...(i)

Also $(f^{-1} og^{-1})(z) = f^{-1}[g^{-1}(z)]$

$= f^{-1}(y)$ [$\because g(y) = z \Rightarrow y = g^{-1}(z)$]

$= x$ [$\because f(x) = y \Rightarrow x = f^{-1}(y)$]

Thus $(f^{-1}og^{-1})(z) = x$ ...(ii)

Hence from (i) and (ii) we have,

$$(gof)^{-1} = (f^{-1} \text{ o } g^{-1}).$$

MISCELLANEOUS SOLVED EXAMPLES

Example 1: *Let the functions $f : R \rightarrow R$, $g : R \rightarrow R$ be defined by*

$f(x) = 2x + 1$, $g(x) = x^2 - 2$.

Find the formulae for g o f and f o g.

Solution: Let us first find the formula for $(g \text{ o } f) : R \rightarrow R$.

We have $(g \text{ o } f)(x) = g[f(x)] = g(2x + 1)$

$$= (2x + 1)^2 - 2 = 4x^2 + 4x - 1.$$

Now let us find the formula for f o g. We have

$(f \circ g)(x) = f[g(x)] = f(x^2 - 2) = 2(x^2 - 2) + 1 = 2x^2 - 3.$

Example 2: *If $f : R \to R$ be defined by $f(x) = |x|$, then prove that $f \circ f = f$.*

Solution: Since f : R → R, therefore the composite mapping f o f is also a mapping from R to R *i.e.*, f o f : R → R.

Now if x be any element of R, we have

$$(f \circ f)(x) = f[f(x)] = f(|x|)$$
$$= ||x|| \quad |x| = f(x).$$

Thus we have $(f \circ f)(x) = f(x), \forall x \in R$.

Hence be the definition of the equality of two mappings,

we have $\quad f \circ f = f.$

Example 3: *Give an example to show that composition of two functions is not in general commutative.*

Solution: Consider the two mappings f and g from R to R given. These mappings f and g are such that $f \circ g \neq g \circ f$.

Hence, the composite of mappings is not a commutative operation.

Example 4 : *If $f : R \to R : f(x) = |x|$, then prove that $f \circ f = f$.*

Solution : $(f \circ f)(x) = f[f(x)] = f(|x|)$

$$= ||x|| = |x| = f(x)$$

$$\therefore \quad f \circ f = f$$

Example 5 : *If the mappings f and g are given by*

$$f = \{(1, 2), (3, 5), (4, 1)\}$$
$$g = \{(2, 3), (5, 1), (1, 3)\}$$

then write down pairs in the mappings (f og) and (g of).

Solution : Here, range (f) = {2, 5, 1} and dom (g) = {2, 5, 1}

∴ Range (f) ⊂ dom (g), so (gof) is defined.

Also, dom (gof = dom(f) = {1, 3, 4}

Now, (gof) (1) = g[f(1)] = g(2) = 3,

(gof) (3) = g[f(3)] = g(5) = 1,

(gof) (4) = g[f(4)] = g(1) = 3.

$\therefore$ gof = {(1, 3), (3, 1), (4, 3)}.

Also, range (g) = {3, 1}, dom (f) = {1, 3, 4}.

Since range (g) $\subset$ dom (f), so (fog) is also defined.

Also dom (fog) = dom (g) = {2, 5, 1}

Now, (fog) (2) = f[g(2)] = f(3) = 5;

(fog) (5) = f[g(5)] = f(1) = 2;

(fog) (1) = f[g(1)] = f(3) = 5.

$\therefore$ fog = {(2, 5), (5, 2), (1, 5)}.

Example 6 : *Decide whether or not the following are functions from A → B where A = {1, 2, 3, 4, 5}, B = {a, b, c, d, e}. If they are functions give the range of each. If they are not, tell why?*

(a) f = {(1, a), (2, b), (3, b), (5, c)}

(b) g = {(1, c), (5, d), (3, a), (2, b), (1, d), (4, a)}

(c) h = {(5, a), (1, c), (4, b), (3, c), (2, d)}

Solution :

(a) Since the element 4 $\in$ A is not associated to any element $\in$ B, therefore f is not a function from A → B.

(b) The element 1 $\in$ A is associated to two different elements c and d $\in$ B. Therefore, g is not a function from A → B.

(c) Each element of A is associated to a unique element of B. Therefore h is a function from A → B. The range of h is the set of the h-images of all elements of A. Thus range of h = h(A) = {a, c, b, c, d} = B.

Example 7: *Let R be the set of all real numbers. Using the fact that every cubic equation with real coefficients has a real root, show that $x \to x^2 - x$ defines a mapping of R onto R. Is this a one-one mapping ?*

Solution: If $x \in R$, then $x^3 - x \in R$ and is unique.

Therefore $x \to x^2 - x$ defines a mapping of R to R.

Let y be any arbitrary element $\in$ R *i.e.*, let y be any real number. Then $x^3 - x = y$ is a cubic equation with real coefficients. It will have at least one real root. Thus for any $y \in R$ there exists $x \in R$ such that $x^3 - x = y$. Therefore $x \to x^2 - x$ defines a mapping of R onto R.

Again $1 \to 1^3 - 1$ *i.e.*, $1 \to 0$

and $-1 \to (-1)^3 - (-1)$

i.e., $-1 \to 0$.

Thus, the two elements 1 and $-1 \in R$ map onto the same element $0 \in R$. Hence the mapping is not one-one.

Example 8: *Let* $X = \{x : x \in R$

and $-\pi/2 \le x \le \pi/2$ *i.e.,*

let $X = [-\pi/2, \pi/2]$

and $Y = (y : y \in R$

and $-1 \le y \le 1\}$ *i.e., let* $Y = [-1, 1]$.

Show that the function $f : X \to Y$ *defined by*

$$f(x) = \sin x\ ,\ (x \in X', \text{ is one-one onto.}$$

Also give the inverse mapp $f^{-1} : Y \to X$.

Solution: Let m and n be any two different real numbers lying in the closed interval $[-\pi/2, \pi/2]$. We know that any two different real numbers lying in the closed interval $[-\pi/2, \pi/2]$ have not the same sine.

$\therefore\ m \ne n \Rightarrow \sin m \ne \sin n$

$\Rightarrow f(m) \ne f(n)$. Hence f is one-one.

Again if y is any arbitrary real number lying in the closed interval $[-1, 1]$, $\exists$ a real number x lying in the closed interval $[-\pi/2, \pi/2]$ such that is n $x = y$.

Thus, every element y in Y is the f-image of some element x in X. Hence f is onto.

Thus $f : X \to Y$ is one-one onto, therefore f has an inverse function $f^{-1} : X \to Y$.

Let y be the image of x under the function f. Then

$$y = f(x) = \sin$$

Consequently, x will be the image of y under the inverse function f^{-1} *i.e.,*

$$x = f^{-1}(y).$$

Solving for x in terms of y in the equation $y = \sin x$, we get

$$x = \sin^{-1} y.$$

Thus $f^{-1}(y) = \sin^{-1} y$, $(y \in Y)$ is the formula defining the inverse function $f^{-1} : Y \to X$.

Example 9 : *Let f be a function defined from the set X to the set Y and let A, B be the subsets of Y, then* *(i)* $f^{-1}(A \cup B) = f^{-1}(A) \cup f^{-1}(B)$

(ii) $f^{-1}(A \cap B) = f^{-1}(A) \cap f^{-1}(B)$.

Example 10: *If* $f : X \to Y$ *and* $A, B \subseteq X$. *Then prove that*

$$f(A \cup B) = f(A) \cup f(B).$$

Solution: Let y be any arbitrary element of $f(A \cup B)$

Then $y \in f(A \cup B)$

$\Rightarrow \exists\ x \in A \cup B$ such that $f(x) = y$

$\Rightarrow \exists\ x \in A$ or $x \in B$ such that $f(x) = y$

$\Rightarrow y \in f(A)$ or $y \in f(B)$

$\Rightarrow y \in f(A) \cup f(B)$.

$\therefore f(A \cup B) \subseteq f(A) \cup f(B)$. ...(1)

Again let y be any arbitrary element of $f(A) \cup f(B)$.

Then $y \in f(A) \cup f(B)$

$\Rightarrow y \in f(A)$ or $y \in f(B)$

$\Rightarrow \exists\ x \in A$ or $x \in B$ such that $f(x) = y$

$\Rightarrow \exists\ x \in A \cup B$ such that $f(x) = y$

$\Rightarrow y \in f(A \cup B)$

$\therefore f(A) \cup f(B) \subseteq f(A \cup B)$. ...(2)

From (1) and (2), we conclude that

$$f(A \cup B) = f(A) \cup f(B).$$

Example 11: *Let* $X = \{x : x \in R$

and $-\pi/2 \le x \le \pi/2$ *i.e.,*

let $X = [-\pi/2, \pi/2]$

and $Y = (y : y \in R$

and $-1 \le y \le 1\}$ *i.e., let* $Y = [-1, 1]$.

Show that the function $f : X \to Y$ *defined by*

$$f(x) = \sin x\ ,\ (x \in X', \text{ is one-one onto.}$$

Also give the inverse mapp $f^{-1} : Y \to X$.

Solution: Let m and n be any two different real numbers lying in the closed interval $[-\pi/2, \pi/2]$. We know that any two different real numbers lying in the closed interval $[-\pi/2, \pi/2]$ have not the same sine.

$\therefore$ m $\neq$ n $\Rightarrow$ sin m $\neq$ sin n

$\Rightarrow$ f (m) $\neq$ f (n). Hence f is one-one.

Again if y is any arbitrary real number lying in the closed interval [–1, 1], $\exists$ a real number x lying in the closed interval [– π/2, π/2] such that is n x = y.

Thus, every element y in Y is the f-image of some element x in X. Hence f is onto.

Thus f : X → Y is one-one onto, therefore f has an inverse function f^{-1} : X → Y.

Let y be the image of x under the function f. Then

$$y = f(x) = \sin$$

Consequently, x will be the image of y under the inverse function f^{-1} *i.e.,*

$$x = f^{-1}(y).$$

Solving for x in terms of y in the equation y = sin x, we get

$$x = \sin^{-1} y.$$

Thus $f^{-1}(y) = \sin^{-1} y, (y \in Y)$ is the formula defining the inverse function f^{-1} : Y → X.

Example 12: *Decide whether or not the following are functions from A to B where A = {1, 2, 3, 4, 5} and B = {a, b, c, d, e}.*

If they are functions, give the range of each. If they are not, tell, why?

1. f = ({1, a), (2, b), (3, b), (5, e)}.
2. g = {(1, e), (5, d), (3, a), (2, b), (1, d), (4, a)}.
3. h = {(5, a), (1, e), (4, b), (3, c), (2, d)}.

Solution:

1. Since the element 4 $\in$ A is not associated to any element $\in$ B, therefore f is not a function from A to B.
2. The element 1 $\in$ A is associated to two different elements e and d $\in$ B. Therefore g is not a function from A to B.
3. Each element of A is associated to a unique element of B. Therefore h is a function from A → B. The range of h is the set of the h-images of all elements of A. So range of h = h (A) = {a, e, b, c, d} = B.

Example 13: *Let A = (–2, –1, 0, 1, 2}. Let the function f : A → R be defined by the promulas f (x) = x^2 + 1. Find the range of f.*

Solution: The range of f consists of those elements of R which appear as f-image of different elements of A. So we calculate the f-image of each element of A.

$$f(-2) = (-2)^2 + 1 = 5,\ f(-1) = (-1)^2 + 1 = 2,$$
$$f(0) = (0)^2 + 1 = 1,\ f(1) = (1)^2 + 1 + 1 = 2.$$
$$f(2) = (2)^2 + 1 = 4 + 1 = 5.$$

Thus, the range of f is the set {5, 2, 1, 2, 5}, *i.e.,* the set {5, 2, 1}.

Example 14: *Let N be the set of natural numbers and A be the set of even natural numbers.*

Let $f : N \to A$ *be defined by the formula* $f(x) = 2x, x \in N$. *Show that the mapping f is one-one onto. Find the formula that defines the inverse function* f^{-1}.

Solution: First to prove that f is a one-one mapping.

Suppose m and n are any two different elements $\in$ N.

Then $m \neq n \Rightarrow 2m \neq 2n \Rightarrow f(m) \neq f(n)$.

Thus different elements belonging to N have different f-images in A. Hence f is one-one.

Now to prove the f is an onto mapping. Let y be any arbitrary element in A *i.e.,* let y be any even natural number. Then f (y/2) = y and y/2 is a natural number *i.e.,* $y/2 \in N$. Thus each element in A is the f-image of some element in N. Hence f is onto.

Since $f : N \to A$ is one-one onto, therefore f has an inverse function $f^{-1} : A \to N$.

Let y be the image of x under the function f. Then

$$y = f(x) = 2x.$$

Consequently, x will be the image of y under the inverse function f^{-1} *i.e.,*

$$x = f^{-1}(y).$$

Solving for x in terms of y in the equation y = 2x, we get

$$x = y/2.$$

Then $$f^{-1}(y) = y/2.$$

Hence $f^{-1}(y) = y/2$, $(y \in A)$ is a formula defining the inverse function $f^{-1} : A \to N$.

Example 15: *Let S be the set of all triangles and* R_+ *be the set of positive real numbers. Prove that the map* $f : S \to R_+$ *given by*

$f(\Delta)$ = area of the Δ, $(\Delta \in S)$, is many-one and onto.

Solution: Let y be any arbitrary element in R_+ *i.e.,* let y be any positive real number. Then $\exists$ a triangle whose area is equal to y. Thus, every element in R_+ is the f-image of some element in S. Therefore f is onto.

Again there can be two or more triangles which have the same area. Therefore f is many-one.

Hence $f : S \to R_+$ is many-one onto.

Example 16: *Let Q be the set of rational numbers. Let $f : Q \to Q$ be defined by*

$$f(x) = 2x + 3, \ (x \in Q).$$

Show that f is one-one and onto. Also find a formula that defines the inverse function f^{-1}.

Solution: Let m and n be any two different elements in Q.

Then
$$m \neq n \Rightarrow 2m \neq 2n$$
$$\Rightarrow 2m + 3 \neq 2n + 3$$
$$\Rightarrow f(m) \neq f(n).$$

Thus, different elements in Q have different f-images in Q. Hence f is one-one.

Let y be any arbitrary element in Q. If y = 2x + 3, we have

$x = (y - 3)/2$ which is also a rational number.

Thus $\left(\dfrac{y-3}{2}\right)$ = y *i.e.,* any arbitrary element y in Q is the f-image of the element $(y - 2)/2 \in Q$. Hence f is onto

Since $f : Q \to Q$ is one-one onto, therefore f has an inverse function $f^{-1} : Q \to Q$.

Let y be the image of x under the function f. Then

$$y = f(x) = 2x + 3.$$

Consequently, x will be the image of y under the inverse function f^{-1} *i.e.,* $x = f^{-1}(y)$.

Solving for x in terms of y in the equation y = 2x + 3, we get

$$x = (y - 3)/2.$$

Thus, $f^{-1}(y) = (y - 3)/2$, $(y \in Q)$ is the formula defining the inverse function $f^{-1} : Q \to Q$.

Note: In order to prove that the mapping f is one-one, we can also argue like this:

Let m and n be any two elements in Q. Then

$f(m) = f(n)$

$\Rightarrow 2m + 3 = 2n + 3$

$\Rightarrow 2m = 2n \Rightarrow m = n.$

Therefore f is one-one.

Example 17 : *Decide whether or not the following are functions from* $A \to B$ *where* $A = \{1, 2, 3, 4, 5\}$, $B = \{a, b, c, d, e\}$. *If they are functions give the range of each. If they are not, tell why?*

(a) $f = \{(1, a), (2, b), (3, b), (5, c)\}$

(b) $g = \{(1, c), (5, d), (3, a), (2, b), (1, d), (4, a)\}$

(c) $h = \{(5, a), (1, c), (4, b), (3, c), (2, d)\}$

Solution :

(a) Since the element $4 \in A$ is not associated to any element $\in B$, therefore f is not a function from $A \to B$.

(b) The element $1 \in A$ is associated to two different elements c and d $\in B$. Therefore, g is not a function from $A \to B$.

(c) Each element of A is associated to a unique element of B. Therefore h is a function from $A \to B$. The range of h is the set of the h-images of all elements of A. Thus range of $h = h(A) = \{a, c, b, c, d\} = B$.

Example 18 : *If* $f : R \to R$ *is defined as follows :*

$$f(x) = \begin{cases} 1, \text{ if } x \in Q \\ -1, \text{ if } x \notin Q \end{cases}$$

Find:

(a) $f\left(\frac{1}{2}\right)$, $f(\pi)$, $f\left(\sqrt{2}\right)$

(b) *Range of f*

(c) *Pre-images of 1 and –1.*

Solution :

(a) It is evident from the definition of f that at every rational point the function attains value 1 and at every irrational point it attains value – 1. So,

$$\frac{1}{2} \in Q \Rightarrow f\left(\frac{1}{2}\right) = 1,\ \pi \notin Q \Rightarrow f(\pi) = -1$$

and $\quad \sqrt{2} \notin Q \Rightarrow f(\sqrt{2}) = -1.$

(b) We have, Range of f = $\{f(x) : x \in R\}$ Also, by definition f(x) attains values 1 and – 1 according as x is rational or irrational and a real number is either rational or irrational.

$\therefore$ Range of f = $\{1, -1\}$.

(c) Since f(x) = 1 for all $x \in Q$. Therefore pre-images of 1 are rational numbers, *i.e.*,

$f^{-1}(1) = Q$.

Also, – 1 is the image of every real number which is not rational. So,

$f^{-1}(-1) = R - Q$ = set of irrational numbers.

Example 19 : *Let A = {– 2, – 1, 0, 1, 2}. Let the function f : A → R is defined by f(x) = $x^2 + 1$.*

Find the range of f.

Solution : The range of f consists of those elements of R which appear as f-images of different elements of A. So, we calculate the f-image of each element of A.

$$f(-2) = (-2)^2 + 1 = 4 + 1 = 5$$
$$f(-1) = (-1)^2 + 1 = 1 + 1 = 2$$
$$f(0) = (0)^2 + 1 = 0 + 1 = 1$$
$$f(1) = (1)^2 + 1 = 1 + 1 = 2$$
$$f(2) = (2)^2 + 1 = 4 + 1 = 5.$$

The range of f is the set {5, 2, 1, 2, 5} *i.e.*, the set {1, 2, 5}.

Example 20 : *Find the domain for which the function f(x) = $2x^2 - 1$ and g(x) = 1 – 3x are equal.*

Solution : We have f(x) = g(x)

$$\Rightarrow \quad 2x^2 - 1 = 1 - 3x$$
$$\Rightarrow \quad 2x^2 + 3x - 2 = 0$$
$$\Rightarrow \quad (x + 2)(2x - 1) = 0$$
$$\Rightarrow \quad x = -2, 1/2.$$

Thus, f(x) and g(x) are equal on the set {–2, 1/2}.

Example 21 : *Is $g = \{(1, 1), (2, 3), (3, 5), (4, 7)\}$ a function? If this is described by the formulae, $g(x) = \alpha x + \beta$, then what values should be assigned to α and β?*

Solution : Since no two ordered pairs in g have the same first component. So, g is a function such that $g(1) = 1$, $g(2) = 3$, $g(3) = 5$ and $g(4) = 7$.

It is given that $g(x) = \alpha x + \beta$. Therefore $g(1) = 1$

and $g(2) = 3 \Rightarrow \alpha + \beta = 1$

and $2\alpha + \beta = 3 \Rightarrow \alpha = 2, \beta = -1$.

Example 22 : *Show that the mapping $f : I \to I$ defined by $f(x) = x^2$, $x \in I$ where I is the set of positive integers, is one-one into.*

Solution : $f(x) = x^2$ means that the function f is such that f-image of x is x^2. Domain of the mapping is $\{1, 2, 3, ...\}$ and the range is $\{1, 4, 9, ...\}$.

Thus f-image is the subset of its domain, *i.e.*, $\{f(x)\} \subset I$. It is a mapping of I into I. Here two different elements of domain necessarily correspond to different elements of the range so that it is one-one mapping. Hence, it is one-one into mapping.

Example 23 : *Show that the function $f : R \to R$ defined by $f(x) = 3x^3 + 5$ for all $x \in R$ is a bijection.*

Solution : *Injectivity* Let x, y be any two elements of R (domain).

Then $$f(x) = f(y) \Rightarrow 3x^3 + 5 = 3y^3 + 5$$

$$\Rightarrow \quad x^3 = y^3 \Rightarrow x = y$$

Thus, $f(x) = f(y) \Rightarrow x = y$ for all $x, y \in R$. So, f is an injective map.

Surjectivity : Let y be an arbitrary elements of R (co-domain). Then

$$f(x) = y \Rightarrow 3x^3 + 5 = y \Rightarrow x^3 = \frac{y-5}{3}$$

$$\Rightarrow \quad x = \left(\frac{y-5}{3}\right)^{1/3}.$$

Thus, we find that for all $y \in R$ (co-domain) there exists

$x = \left(\frac{y-5}{3}\right)^{1/3} \in R$ (domain) such that

$$f(x) = f\left[\left(\frac{y-5}{3}\right)^{1/3}\right] = 3\left[\left(\frac{y-5}{3}\right)^{1/3}\right]^3 + 5 = y - 5 + 5 = y.$$

This shows that every element in the co-domain has its pre-image in the domain. So, f is a surjection.

Hence, f is a bijection.

Example 24 : *Let* $A = \{x \in R \mid -1 \leq x \leq 1\} = B$. *Show that* $f : A \to B$ *given by* $f(x) = x|x|$ *is a bijection.*

Solution : *Injectivity :* Let x, y be any two elements in A. Then

$$x \neq y \Rightarrow x|x| \neq y|y| \Rightarrow f(x) \neq f(y)$$

So, $f : A \to B$ is an injective map.

Surjectivity : Clearly, range of f = f(A) = B. So, $f : A \to B$ is a surjective map. Hence, f is a bijection.

Example 25 : *If R is the set of real numbers, discuss the mapping*

$$f : R \to R$$

where $f(x) = x^2, x \in R$.

Solution : If x is any real number then x^2 is also a real number and it will be unique. Thus each element $x \in R$ has a unique f-image in R. Therefore, $f : R \to R$.

Since, there is no real number whose square is negative, therefore any negative number in R is not the f-image of any element in R. Consequently f is mapping of R into R.

Also, we see that $(-2)^2 = 4$ and $(2)^2 = 4$. Thus the elements 2 and -2 in R have the same f-image 4 in R. Hence, f is many-one into mapping of $R \to R$.

Example 26 : *Let C and R denote the set of all complex numbers and all real numbers respectively. Then show that* $f : C \to R$ *given by* $f(z) = |z|$ *for all* $z \in C$ *is neither one-one nor onto.*

Solution : *Injectivity :* We find that $z_1 = 1 - i$ and $z_2 = 1 + i$ are two distinct complex numbers such that $|z_1| = |z_2|$, *i.e.*, $z_1 \neq z_2$ but $f(z_1) = f(z_2)$.

This shows that different elements may have the same image. So, f is not an injection, *i.e.*, not one-one. *Surjectivity :* f is not a surjection, because negative real numbers in R do not have their pre-images in C. In other words, for every negative real number a there is no complex number $z \in C$ such that $f(z) = |z| = a$. So, f is not a surjection, *i.e.*, not onto.

Hence, $f : C \to R$ is neither one-one nor onto.

Example 27 : *Let* $A = \{0, 1\}$ *and let N be the set of natural numbers. Show that* $f : N \to A$ *defined by* $f(2x) = 0, f(2x + 1) = 1, (x \in N)$ *is a many-one onto mapping of N to A.*

Solution : Here we see that each element in A is the f-image of at least one element in N. For example 0 is the f-image of $4 \in N$ and 1 is the f-image of $9 \in N$. Therefore, f is a mapping of N onto A.

Also, we see that the element $1 \in A$ is the f-image of every odd natural number $\in N$. Hence, f is a many-one onto mapping of N to A.

Example 28: *Let R_0 be the set of all non-zero real numbers. Let $f : R_0 \to R_0$ be defined by the formula $f(x) = 1/x$, $x \in R_0$. Show that f is one-one and onto mapping.*

Solution: Let m and n be any two different elements in R_0.

Then $\quad m \neq n \Rightarrow 1/m \neq 1/n \Rightarrow f(m) \neq f(n)$.

Thus, different elements in R_0 have different f-images in R_0. Hence f is one-one.

Now let y be an arbitrary element in R_0. If y 1/x, we have $x = 1/y$ which is a non-zero real number if y is a non-zero real number. Thus $f(1/y) = y$ *i.e.,* any arbitrary element y in R_0 is the f-image of the element $1/y$ in R_0. Hence f is onto.

Thus $\quad f : R_0 \to R_0$ is one-one onto.

It should be noted that $f(x) = 1/x$ cannot be a mapping from the set of real numbers to the set of real numbers. Then reason is that the image of 0 will not exist because $f(0) = 1/0$ and $1/0$ is not a real number.

Example 29: *Let C be the set of complex numbers. Prove that the map $f : C \to R$ given by $f(z) = |z|$, $z \in C$*

is neither one-one nor onto.

Solution: Let $z = x + iy$ be any complex number, where $x, y \in R$ and i $\sqrt{(-1)}$.

Then $|z| = \sqrt{(x^2 + y^2)}$ Also $|z|$ is always a non-negative real number *i.e.,* $|z| \geq 0$.

If m be any negative real number $\in R$, then there exists no complex number $z \in C$ such that $|z| = m$. Thus m is not the f-image of any complex number $\in C$. Hence f is not onto but is into.

Also $\quad z_1 = 2 + i3 \in C$ and $z_2 = 2 - i3 \in C$.

Then $\quad |2 + i3| = \sqrt{13}$ and $|2 - i3| = \sqrt{13}$.

Thus $\quad z_1 \neq z_2$, although $|z_1| = |z_2|$ *i.e.,* $f(z_1) = f(z_2)$.

Thus we see that two different complex numbers z_1 and $z_2 \in C$ have the same f-image in R. Hence f is not one-one.

Example 30: *Show that the mapping* $f : I_+ \to I_+$ *defined by* $f(x) = x^2$, $x \in I_+$ *where* I_+ *is the set of positive integers, is one-one into.*

Solution: It is given that f is a mapping from I_+ to I_+ defined by

$$f(x) = x^2 \ \forall \ x \in I_+.$$

We have $\quad I_+ = \{1, 2, 3, 4, ...\}$.

f is one-one. Let x_1 $x_2 \in I_+$. We have

$$f(x_1) = f(x_2) \Rightarrow x_1^2 = x_2^2 \text{ [by def. of f]}$$

$$\Rightarrow x_1 = x_2 \Rightarrow \text{f is one-one.}$$

f is into. Now we shall prove that f is into and not onto. We have $2 \in I_+$. Now there exists no $x \in I_+$ such that $x^2 = 2$. Thus $2 \in I_+$ is not the f-image of any $x \in I_+$. Therefore the mapping f is into and not onto.

Example 31: *Classify the following as mappings of* $R \to R$:

(a) $f(x) = \log x$,

(b) $f(x) = + \sqrt{x}$.

(c) $f(x) = \tan x$, $(x \in R)$.

Solution: (a) Let $x \in R$ and let x be a negative real number. Then log x is not a real number, *i.e.*, log x is not an element of R. Thus the negative real numbers in R will have no f-image in R. Also $\log 0 = -\infty$ which is not an element of R. Thus, the real number 0 in R has no f-image in R. Hence f is not a function from R to R. However f is a function from R_+ to R.

(b) Let x be a negative real number. Then $\sqrt{x}$ is not a real number *i.e.*, $\sqrt{x} \notin R$. Thus the negative real numbers in R will have no f-images in R. Hence f is not a function from R to R. If S denotes the set of all non-negative real numbers, then f defined by this formula is a function from S to R.

(c) Let $x = \pi/2$. Then $\tan \pi/2 =$ which is not an element of R. Thus $\pi/2 \in R$ has no f-image in R. In general if $x \in R$ is such that $x = n.\frac{\pi}{2}$ Where n is an odd integer, then x has no f-image in R. Hence f is not a function from R to R.

However, if $X = \{x : x \in R \text{ and } -\pi/2 < x < \pi/2\}$, then f is a function from X to R, Obviously this function will be one-one and onto.

Example 32: *If R is the set of real numbers discuss the mapping*

$f : R \to R$ *where* $f(x) = x^2$, $x \in R$

What is the domain and the range of the mapping f ?

Solution: If x is any real number, the x^2 is also a real number and it will be unique. Thus each element $x \in R$ has a unique f-image in R. Therefore $f : R \to R$.

Since there is no real number whose square is negative, therefore any negative number in R is not the f-image of any element in R. Consequently f is a mapping of R into R and not from R onto R.

Also we see that $(2)^2 = 4$ and $(-2)^2 = 4$. Thus the elements 2 and –2 in R have the same f-image 4 in R. Hence f is a many-one into mapping of R to R.

Domain of f: Since $f : R \to R$, therefore domain of f is R.

Range of f: If $x \in R$, then $f(x) = x^2$ is a non-negative real number. Further if y is any non-negative real number, then $\exists$ a real number $\sqrt{y}$ such that $f(\sqrt{y}) = (\sqrt{y})^2 = y$. Thus, every non-negative real number is the f-image of some real number. Therefore the range of f is the set of non-negative real numbers.

Example 33: *Let A = { a, b, c,} and B = {e, f}. Write down all possible functions from A to B.*

Solution: While forming a mapping from A to B, each element of A can be associated to an element of B in two ways. Thus we can have $2 \times 2 \times 2$ *i.e.,* 8 distinct mappings from A to B.

These are as given below:

{(a, e), (b, e), (c, e)}, {(a, e), (b, e), (c, f)},
{(a, e), (b, f), (c, e)}, {(a, e), (b, f), (c, f)},
{(a, f), (b, e), (c, e)}, {(a, f), (b, e), (c, f)},
{(a, f), (b, f), (c, e)}, {(a, f), (b, f), (c, f)}.

Example 34: *Each of the following formulas defines a function from R to R. Find the range of each function.*

1. $f(x) = x^3$;
2. $g(x) = \sin x$;
3. $h(x) = x^2 + 1$.

Solution:

1. We know that every real number a has a real cube root $\left(\sqrt[3]{a}\right)$.

   Therefore if a be any arbitrary element $\in R$, then $f\left(\sqrt[3]{a}\right) = \left(\sqrt[3]{a}\right)^3$

   $= a$ *i.e.,* $\sqrt[3]{a} \in R$ is the pre-image of $a \in R$. Since a is any arbitrary element $\in$ R, therefore the range of f is R. This function will be an onto function.

2. The sine of any real number lies in the closed interval [–1, 1]. Also all the numbers in this interval will be the sine of some real numbers. Hence the range of g is the closed interval [–1, 1]. This function will be an into function.
3. If we add 1 to the square of each real number, we get the set of numbers which are greater then or equal to 1. Let y be any real number greater than or equal to 1 and let $y = h(x) = x^2 + 1$.

   Then $x = \pm \sqrt{(y-1)}$ are also real numbers.

Thus, every real number which is greater than or equal to 1 is the h-image of some or other real number. Hence the range of h is the infinite interval $(1, \infty)$.

Example 35: *Show that the mapping $f : R \to R$ defined by*

$$f(x) = \cos x \ \forall x \in R$$

is neither one-one nor onto. Modify the domain and co-domain of this mapping so that it may be both one-to-one and onto.

Solution: We have $f : R \to R$ such that

$$f(x) = \cos x \ \forall \ x \in R.$$

f is not one-one. We have $0, 2\pi \in R$

and $f(0) = \cos 0 = 1$, $f(2\pi) = \cos 2\pi = 1$.

Thus, 0 and 2π are two distinct elements in the domain R of f and they have the same f-image 1 in the co-domain R of f. Therefore f is may-one and not one-one.

f is not onto: We know that if $x \in R$, then $-1 \le \cos x \le 1$. Therefore if $y \in$ the co-domain R of f and $y > 1$ or $y < -1$, then there exists no $x \in$ the domain R of f such that $f(x) = \cos x = y$. Therefore the mapping f is into and not onto.

Now let $X = \{x : x \in R \text{ and } 0 \le x \le \pi\}$

and $\quad Y = \{y : y \in R \text{ and } -1 \le y \le 1\}$.

If we take X as the domain of f and Y as the co-domain of f, then the mapping $f : X \to Y$ such that

$$f(x) = \cos x \ \forall \ x \in X$$

is one-one and onto.

Example 36: *Let R be the set of all real numbers. Using the fact that every cubic equation with real coefficients has a real root, show that* $x \to x^2 - x$ *defines a mapping of R onto R. Is this a one-one mapping ?*

Solution: If $x \in R$, then $x^3 - x \in R$ and is unique.

Therefore $x \to x^2 - x$ defines a mapping of R to R.

Let y be any arbitrary element $\in$ R *i.e.,* let y be any real number. Then $x^3 - x = y$ is a cubic equation with real coefficients. It will have at least one real root. Thus for any $y \in R$ there exists $x \in R$ such that $x^3 - x = y$. Therefore $x \to x^2 - x$ defines a mapping of R onto R.

Again $\quad 1 \to 1^3 - 1$ *i.e.,* $1 \to 0$

and $\quad -1 \to (-1)^3 - (-1)$

i.e., $\quad -1 \to 0.$

Thus, the two elements 1 and $-1 \in R$ map onto the same element $0 \in R$. Hence the mapping is not one-one.

Example 37:

If S and T are non-empty sets, prove that there exists a one-to-one correspondence between $S \times T$ *and* $T \times S$.

Solution:

If $a \in S$ and $b \in T$, then the ordered pair $(a, b) \in S \times T$ and the ordered pair $(b, a) \in T \times S$. Let f be a mapping from $S \times T$ to $T \times S$ defined by the formula

$$f(a, b) = (b, a) \ \forall \ (a, b) \in S \times T.$$

Here f (a, b) denotes the image under the mapping f of the element $(a, b) \in S \times T$.

Obviously the mapping f is well-defined.

f is one-one. Let $(a, b), (c, d) \in S \times T$. Then $f(a, b) = f(c, d)$

$\Rightarrow \quad (b, a) = (d, c)$ $\quad$ [by def. of f]

$\Rightarrow \quad b = d, a = c$

$\Rightarrow \quad (a, b) = (c, d)$

$\Rightarrow \quad$ f is one-one.

f is onto: Let (y, z) be any element of $T \times S$. Then $y \in T, z \in S$ Therefore $(z, y) \in S \times T$. We have $f(z, y) = (y, z)$.

Thus, $(y, z) \ T \times S \Rightarrow \exists \ (z, y) \in S \times T$ such that $f(z, y) = (y, z)$. Therefore f is onto.

Thus, f is a one-one function from $S \times T$ onto $T \times S$. Therefore f gives a one-to-one correspondence between $S \times T$ and $T \times S$.

Example 38 : *If the mapping $f : I \to I$ be defined by $f(x) = x^2$ where I is the set of integers, evaluate $f^{-1}(16)$ and $f^{-1}(-3)$. Does the inverse mapping exist? If so, find it.*

Solution : Let $f^{-1}(16) = x$ where $x \in I$

$$16 = f(x) = x^2$$

$$\therefore \quad x = 4 \text{ or } -4$$

Hence, $f^{-1}(16) = \{-4, 4\}$

Also, since there is no integer whose square is -1, it follows that $f^{-1}(-3) = \phi$, the empty set.

Under the given mapping, the elements -4 and 4 have the same image 16. Accordingly, the given mapping is not one-one. Hence the inverse mapping does not exist.

Example 39: *If S and T are non-empty sets, prove that there exists a one-to-one correspondence between $S \times T$ and $T \times S$.*

Solution : (a) Let x be an element of $f^{-1}(A \cup B)$, then

$$x \in f^{-1}(A \cup B)$$

$$\Rightarrow \quad f(x) \in A \cup B$$

$$\Rightarrow \quad f(x) \in A \text{ or } f(x) \in B$$

$$\Rightarrow \quad x \in f^{-1}(A) \text{ or } x \in f^{-1}(B)$$

$$\Rightarrow \quad x \in \{f^{-1}(A) \cup f^{-1}(B)\}$$

$$\therefore \quad f^{-1}(A \cup B) \subset \{f^{-1}(A) \cup \{f^{-1}(B)\} \qquad ...(1)$$

Again, let y be an element of $f^{-1}(A) \cup f^{-1}(B)$, then

$$y \in f^{-1}(A) \cup f^{-1}(B)$$

$$\Rightarrow \quad y \in f^{-1}(A) \text{ or } y \in f^{-1}(B)$$

$$\Rightarrow \quad f(y) \in A \text{ or } f(y) \in B$$

$$\Rightarrow \quad f(y) \in A \cup B$$

$$\Rightarrow \quad y \in f^{-1}(A \cup B)$$

Therefore $f^{-1}(A) \cup f^{-1}(B) \subset f^{-1}(A \cup B)$ ...(2)

(1) and (2), imply.

$$f^{-1}(A \cup B) = f^{-1}(A) \cup f^{-1}(B).$$

Similarly, 2nd result can be proved.

Example 40 : *If the map $f : R \to R$ be given by $f(x) = 4x - 1$ and the map $g : R \to R$ be given by $g(x) = x^3 + 2$, find $(g \circ f)\, x$ and $(f \circ g)\, x$, R being the set of real numbers.*

Solution : We have,

$$(gof)x = g[f(x)] = g(4x - 1) \text{ as } f(x) = 4x - 1$$
$$= (4x - 1)^3 + 2 \text{ as } g(x) = x^3 + 2$$
$$= 64x^3 - 4x^2 + 12x + 1.$$

Also
$$(fog)x = f[g(x)] = f(x^3 + 2)$$
$$= 4(x^3 + 2) - 1$$
$$= 4x^3 + 7.$$

Example 41: *Let the functions $f : R \to R$ and $g : R \to R$ be defined by $f(x) = 2x$, $g(x) = x^2 + 2$, $\forall x \in R$. Check the functions f and g for being*

(i) one-to-one

(ii) onto.

Find the formulae defining the functions $f \circ g$ and $g \circ g$ and obtain the values of $(f \circ g)(2)$ and $(g \circ g)(1)$.

Solution: The mapping f is both one-to-one and onto as shown below.

f is one-to-one: Let $x_1, x_2 \in R$. Then

$$f(x_1) = f(x_2)$$
$$\Rightarrow 2x_1 = 2x_2 \Rightarrow x_1 = x_2.$$

$\therefore$ The mapping f is one-to-one.

f is onto: Let y be any element of R. Then y/2 is also an element of R and we have

$$f(y/2) = 2.(y/2) = y.$$

Thus, each element in R is the f-image of some element in R. Hence the mapping f is onto.

The mapping g is neither one-to-one nor onto as shown below.

g is not one-to-one: We have $1 \in R$ and $-1 \in R$. By the definition of the mapping g, we have

$$g(1) = 1^2 + 2 = 3,$$
$$g(-1) = (-1)^2 + 2 = 3.$$

Thus, 1 and -1 are two distinct elements in R and we have $g(1) = g(-1)$. Therefore the mapping g is not one-to-one.

The mapping g is not onto: We have $-5 \in R$ and there exists no $x \in R$ such that $g(x) = x^2 + 2 = -5$. Therefore the mapping g is not onto.

To find the formula for the mapping $f \circ g : R \to R$.

If $x \in R$, then we have

$$(f \circ g)(x) = f[g(x)] = f(x^2 + 2) = 2(x^2 + 2)$$
$$= 2x^2 + 4.$$

$\therefore$ $f \circ g : R \to R$ is defined by the formula

$$(f \circ g)(x) = 2x^2 + 4, \ \forall \ x \in R.$$

Using this formula, we have

$$(f \circ g)(2) = 2.(2)^2 + 4 = 12.$$

Now to find the formula defining the mapping $g \circ g : R \to R$.

If $x \in R$, then we have

$$(g \circ g)(x) = g[g(x)] = g(x^2 + 2)$$
$$= (x^2 + 2)^2 + 2 = x^4 + 4x^2 + 6.$$

$\therefore$ $g \circ g : R \to R$ is defined as

$$(g \circ g)(x) = x^4 + 4x^2 + 6, \ \forall \ x \in R.$$
$$(g \circ g)(1) = 1^4 + 4.1^2 + 6 = 11.$$

Example 42: *If $f : X \to Y$ and $g : Y \to Z$ are any two mappings, show that if $g \circ f : Y \to Z$ is one-to-one, then f is one-to-one.*

Solution: Let x_1, x_2 be any two elements of X.

Then $f(x_1) = f(x_2)$

$\Rightarrow g[f(x_1)] = g[f(x_2)]$

$\Rightarrow (g \circ f)(x_1) = (g \circ f)(x_2)$

$\Rightarrow x_1 = x_2$, because the mapping $g \circ f$ is given to be one-to-one.

Hence the mapping f is one-to-one.

Example 43: *Let $f : R \to R$ be given by the formula $f(x) = x^2$, $(x \in R)$ and $g : R \to R$ be give by the formula $g(x) = x + 3$, $(x \in R)$. Calculate $(f \circ g)(2)$ and $(g \circ f)(2)$. Hence show that*

$$f \circ g \neq g \circ f.$$

Solution: We have $(f \circ g)(2) = f[g(2)] = f(2 + 3)$

$$-f(5) = 5^2 = 25.$$

Also $(g \circ f)(2) = g[f(2)] = g(2^2)$

$$= g(4) = 4 + 3 = 7.$$

Now both f o g and g o f are mappings from R to R. Since $2 \in R$ is such that $(f \circ g)(2) \neq (g \circ f)(2)$, therefore the mappings f o g and g o f are not equal. Hence $f \circ g \neq g \circ f$.

Example 44: *Let* $A = \{1, 2, 3, 4, 5\}$ *and let the functions*

$$f : A \to A,\ g : A \to A$$

be defined as

$f(1) = 3, f(2) = 5, f(3) = 3,$

$f(4) = 1, f(5) = 2,$

$g(1) = 4, g(2) = 1,$

$g(3) = 1, g(4) = 2,$

$g(5) = 3.$

Find the composite functions $g \circ f$ *and* $f \circ g$.

Solution: We have,

$$(g \circ f)(1) = g[f(1)] = g(3) = 1$$
$$(g \circ f)(2) = g[f(2)] = g(5) = 3$$
$$(g \circ f)(3) = g[f(3)] = g(3) = 1$$
$$(g \circ f)(4) = g[f(4)] = g(1) = 4$$
$$(g \circ f)(5) = g[f(5)] = g(2) = 1.$$

Also
$$(f \circ g)(1) = f[g(1)] = f(4) = 1$$
$$(f \circ g)(2) = f[g(2)] = f(1) = 3$$
$$(f \circ g)(3) = f[g(3)] = f(1) = 3$$
$$(f \circ g)(4) = f[g(4)] = f(2) = 5$$
$$(f \circ g)(5) = f[g(5)] = f(3) = 3.$$

Example 45 : *If* $f : R \to R$ *is defined as follows :*

$$f(x) = \begin{cases} 1, \text{ if } x \in Q \\ -1, \text{ if } x \notin Q \end{cases}$$

Find:

(a) $f\left(\frac{1}{2}\right)$, $f(\pi)$, $f\left(\sqrt{2}\right)$

(b) *Range of f*

(c) *Pre-images of 1 and –1.*

Solution :

(a) It is evident from the definition of f that at every rational point the function attains value 1 and at every irrational point it attains value – 1. So,

$$\frac{1}{2} \in Q \Rightarrow f\left(\frac{1}{2}\right) = 1, \pi \notin Q \Rightarrow f(\pi) = -1$$

and $$\sqrt{2} \notin Q \Rightarrow f\left(\sqrt{2}\right) = -1.$$

(b) We have, Range of f = {f(x) : x ∈ R} Also, by definition f(x) attains values 1 and – 1 according as x is rational or irrational and a real number is either rational or irrational.

∴ Range of f = {1, – 1}.

(c) Since f(x) = 1 for all x ∈ Q. Therefore pre-images of 1 are rational numbers, *i.e.*,

$f^{-1}(1) = Q.$

Also, – 1 is the image of every real number which is not rational. So,

$f^{-1}(-1) = R - Q$ = set of irrational numbers.

Example 46: *Let C be the set of complex numbers. Prove that the map $f : C \to R$ given by $f(z) = |z|, z \in C$*

is neither one-one nor onto.

Solution: Let z = x + iy be any complex number, where x, y ∈ R and i √(–1).

Then $| z | = \sqrt{(x^2 + y^2)}$ Also | z | is always a non-negative real number *i.e.*, $| z | \geq 0$.

If m be any negative real number ∈ R, then there exists no complex number z ∈ C such that | z | = m. Thus m is not the f-image of any complex number ∈ C. Hence f is not onto but is into.

Also $z_1 = 2 + i3 \in C$ and $z_2 = 2 - i3 \in C$.

Then $| 2 + i3 | = \sqrt{13}$ and $| 2 - i3 | = \sqrt{13}$.

Thus $z_1 \neq z_2$, although $| z_1 | = | z_2 |$ *i.e.*, $f(z_1) = f(z_2)$.

Thus we see that two different complex numbers z_1 and $z_2 \in C$ have the same f-image in R. Hence f is not one-one.

Example 47: *Show that the mapping* $f : I_+ \rightarrow I_+$ *defined by* $f(x) = x^2$, $x \in I_+$ *where* I_+ *is the set of positive integers, is one-one into.*

Solution: It is given that f is a mapping from I_+ to I_+ defined by

$$f(x) = x^2 \ \forall \ x \in I_+.$$

We have $\quad I_+ = \{1, 2, 3, 4, ...\}$.

f is one-one. Let $x_1\ x_2 \in I_+$. We have

$$f(x_1) = f(x_2) \Rightarrow x_1^2 = x_2^2 \text{ [by def. of f]}$$

$$\Rightarrow x_1 = x_2 \Rightarrow \text{f is one-one.}$$

f is into. Now we shall prove that f is into and not onto. We have $2 \in I_+$. Now there exists no $x \in I_+$ such that $x^2 = 2$. Thus $2 \in I_+$ is not the f-image of any $x \in I_+$. Therefore the mapping f is into and not onto.

Example 48: *Let* $R \rightarrow R$ *defined by* $f(x) = ax + b$, *where* $a, b, x \in R$ *and* $a \neq 0$. *Prove that f is invertible.*

Solution : f is one-one for

$$x_1, x_2 \in R,\ f(x_1) = f(x_2)$$

$$\Rightarrow \quad ax_1 + b = ax_2 + b \Rightarrow ax_1 = ax_2$$

$$\Rightarrow \quad x_1 = x_2.$$

f is onto :

Let $y \in R$ such that

$$y = f(x) \Rightarrow y = ax + b$$

$$\Rightarrow \quad ax = y - b \text{ and } a \neq 0 \in R$$

$$\Rightarrow \quad x = \frac{1}{a}(y - b) \in R$$

$\therefore$ Given $y \in R$, $\exists$ some $x = \frac{1}{a}(y - b) \in R$ s.t. $f(x) = y$

$\therefore$ $f : R \rightarrow R$ is both one-one and onto hence f is invertible.

Example 49: *Let S be a non-empty set and* $*$ *be a binary operation on S defined by* $x * y = x : x, y \in S$.

Determine whether $*$ *is commutative and associative.*

Solution: Let $x, y \in S$. Then

$$x * y = x \text{ and } y * x = y.$$

Thus if x and y are distinct elements of S, we have

$$x * y \neq y * x.$$

$\therefore$ the operation on S defined by $*$ is not commutative.

Again let x, y, z be any three elements of S. Then

$$(x * y) * z = x * z = x$$

and $x * (y * z) = x * y = x$.

Thus $(x * y) * z = x * (y * z) \ \forall \ x, y, z \in S$.

Hence the operation on S defined by $*$ is associative.

Example 50: *Find whether the following algebraic structures have identity element or not:*

1. (N, +) 2. (N, ·) 3. (I, +); 4. (I, ·); 5. (R, ·);
6. (P (S), ∪); 7. (P (S), ∩).

Solution:

1. The algebraic structure (N, +) does not possess identity element. For ordinary addition of numbers, the number 0 is identity and $0 \notin N$.
2. The algebraic structure (N, ·) possesses identity element and it is 1. We have $1 \in N$ and

$$1a = a = a1 \ \forall \ a \in N.$$

3. The algebraic structure (I, +) possesses identity element and it is 0. We have $0 \in I$ and

$$0 + a = a = a + 0 \quad \forall \ a \in I.$$

4. The algebraic structure (I, ·) possesses identity element and it is 1. We have $1 \in I$ and $1a = a = a1 \ \forall \ a \in I$.
5. The algebraic structure (R, ·) possesses identity element and it is 1.
6. The algebraic structure (P (S), ∪) possesses identity element and it is the null set ∅. We have

   $\varnothing \in P(S)$ because $\varnothing \subseteq S$.

   Also $\quad \varnothing \cup A = A = A \cup \varnothing, \ \forall \ A \in P(S)$.
7. The algebraic structure (P (S), ∩) possesses identity element and it is the set S itself. We have

$$S \in P(S) \text{ because } S \subseteq S.$$

   Also $\quad S \cap A = A = A \cap S, \ \forall \ A \in P(S)$ *i.e.*, $\forall \ A \subseteq S$

Example 51: *Define binary operation. Show that the relation* o *given by* $a \circ b = a^2$ *is a binary operation on the set of natural numbers. Is this binary operation associative.*

Solution: Let a, b, $\in$ N where N is the set of natural numbers. Then a^b is a unique natural numbers. Thus,

$$a \text{ o } b = a^b \in N, \forall a, b \in N.$$

$\therefore$ 'o' is a binary operation on the set N.

Now let a, b, c, $\in$ N. Then

$$(a \text{ o } b) \text{ o } c = (a^b) \text{ o } c = (a^b) = a^{bc} \quad ...(1)$$

and $a \text{ o } (b \text{ o } c) = a \text{ o } (b^c) = (a)^{bc}$ ...(2)

From (1) and (2), we observe that it is not necessary that a o (b o c) = (a o b) o c $\forall$ a, b, c $\in$ N. Hence, the operation on N defined by 'o' is not associative.

For a concrete example, take 2, 2, 3 $\in$ N.

We have $(2 \text{ o } 2) \text{ o } 3 = (2) \text{ o } 3 = 4 \text{ o } 3 = 4^3 = 64$

and $2 \text{ o } (2 \text{ o } 3) = 2 \text{ o } (2^3) = 2 \text{ o } 8 = 2^8 = 256.$

Thus $(2 \text{ o } 2) \text{ o } 3 \neq 2 \text{ o } (2 \text{ o } 3).$

Example 52: *Consider the binary operation $*: Q \times Q \to Q$ which 1 defined by $a * b = a + b + ab$, for every a, b, $\in Q$*

1. *Is $*$ commutative?*
2. *Is $*$ associative?*
3. *Find the identity element for $*$.*
4. *Do any of the elements in Q have an inverse and what is it?*

Solution:

1. We have $a * b = a + b + ab$

   and $b * a = b + a + ba.$

   But $a + b = b + a$ and $ab = ba.$

$\therefore$ $a * b = b * a$ and thus $*$ is commutative.

2. We have $a * (b * c) = a * (b + c + bc)$

   $= a + (b + c + bc) + a (b + c + bc)$

   $= a + b + c + bc + ab + ac + abc.$

   Also $(a * b) * c = (a + b + ab) * c$

   $= (a + b + ab) + c + (a + b + ab) c$

   $= a + b + c + ab + ac + bc + abc.$

$\therefore$ $a * (b * c) = (a * b) * c$ and so $*$ is associative.

3. Let e be the identity element for $*$. Then for every $a \in Q$, we must have $a * e = e * a = a$.

   But $*$ is commutative, therefore $a * e = e * a$.

   Now $a * e = a$, $(\forall\ a \in Q)$

   $\Rightarrow a + e + ae = a \Rightarrow e + ae = 0 \Rightarrow e\ (1 + a) = 0$

   $\Rightarrow e = 0$, since $1 + a \neq 0$, for every $a \in Q$.

   Now $0 \in Q$ and we have

   $0 * a = 0 + a + 0.a = a = a * 0\ \forall\ a \in Q$.

   Hence 0 is the identity element.

4. Since 0 is the identity element, therefore for a to have an inverse b, we must have $a * b = 0$.
   Now $a * b = 0 \Rightarrow a + b + ab = 0 \Rightarrow a + b\ (a + 1) = 0$

   $\Rightarrow a = -\ b\ (a + 1) \Rightarrow b = -\dfrac{a}{a+1}$, if $a \neq -1$.

   Thus if $a \neq -1$, then a has an inverse and it is $-a/(a + 1)$.

Example 53 : *If I the set of integers then show that relation R in I such that a R b iff a – b is divisible by m (positive integer) is an equivalence relation.*

Solution : (i) *Reflexivity* : For each $a \in I$, $a = a = 0$ which is divisible by m, so a R a Therefore, R is reflexive.

(ii) *Symmetry* : Let a R b for $a, b, \in I$,

i.e., $a - b$ is divisible by m. Then as $a - b = -\ (b - a)$

$(b - a)$ is also divisible by m.

Hence $a R b \Rightarrow b R a$ so that R is symmetric.

(iii) *Transitivity* : For $a, b, c \in I$, Let a R b, b R c, *i.e.*, let $(a - b)$, $(b - c)$ be divisible by m. Then as

$(a - b) + (b - c) = a - c$,

$a - c$ is also divisible by m.

Hence, a R b, b R c $\Rightarrow$ a R c, *i.e.*, R is transitive, showing that R is an equivalence relation.

Example 54: *Prove that the relation R in the set of integers I defined by a R b if a and b are both odd is symmetric and transitive but not reflexive.*

Solution:

1. *R is not Reflexive:* We have $2 \in I$. But 2 is not R-related to 2, since 2 and 2 are not both odd. Therefore R is not reflexive.

2. *R is Symmetric:* Suppose we have a R b. Then a and b are both odd. Now if a and b are both odd then b and a are both odd *i.e.,* b R a. Thus a R b

   $\Rightarrow$ b R a and therefore R is symmetric.

3. *R is Transitive:* Suppose we have a R b and b R c. Then a and b are both odd as well as b and c are both odd. It implies that a and c are both odd *i.e.,* a R c. Thus a R b and b R c

   $\Rightarrow$ a R c. Therefore R is transitive.

Example 55: *Show with the help of an example that the union of two transitive relations is not necessarily a transitive relation.*

Solution: Let $R_1 = \{(1, 2), (2, 1), (1, 1), (2, 2)\}$. $R_2\{(2, 3), (3, 2), (2, 2), (3, 3)\}$ be two relations on the set $S = \{1, 2, 3\}$. Obviously both R_1 and R_2 are transitive relations. Now $R_1 \cup R_2 = \{(1, 2), (2, 1), (1, 1), (2, 2), (2, 3), (3, 2), (3, 2), (2, 2), (3, 3)\}$.

This is not a transitive relation because $(1, 2) \in R_1 \cup R_2$ and $(2, 3) \in R_1 \cup R_2$ while $(1, 3) \notin R_1 \cup R_2$.

Example 56: *Prove that the relation "congruence modulo m" is an equivalence relation in the set of integers.*

Solution: Let I be the set of integers. If m is any given positive integer and a, b $\in$ I, then we say that $a \equiv b \pmod{m}$ if $m \mid (a - b)$ *i.e.,* if m is a divisor of a – b. We shall prove that this defines an equivalence relation in the set I.

Reflexivity: Let a be any integer. Then $a - a = 0$ and $m \mid 0$ because we can write $0 = m. 0$. Thus $a \equiv a \pmod{m}\ \forall\ a \in I$. Therefore the relation is reflexive.

Symmetry: Let a, b $\in$ I be such that $a \equiv b \pmod{m}$.

Then $a \equiv b \pmod{m}$

$\Rightarrow m \mid (a - b)$

$\Rightarrow a - b = km$, where k is some integer

$\Rightarrow b - a = (-k)\ m$, where –k is also an integer

$\Rightarrow m \mid (b - a)$

$\Rightarrow b \equiv a \pmod{m}$.

Thus, $a \equiv b \pmod{m}$

$\Rightarrow b \equiv a \pmod{m}$. Therefore the relation is symmetric.

Transitivity: Let a, b, c $\in$ I be such that

$a \equiv b \pmod{m}$, $b \equiv a \pmod{m}$. Then we have

$m \mid (a - b)$ and $m \mid (b - c)$

$\Rightarrow a - b = k_1 m$ and $b - c = k_2 m$, where k_1, k_2 are some integers

$\Rightarrow (a - b) + (b - c) = k_1 m + k_2 m$

$\Rightarrow a - c = (k_1 + k_2)m$, where $k_1 + k_2$ is also an integer

$\Rightarrow m \mid (a - c) \Rightarrow a \equiv c \pmod{m}$.

Thus $a \equiv b \pmod{m}$ and $b \equiv c \pmod{m}$

$\Rightarrow a \equiv c \pmod{m}$. Therefore the relation is transitive.

Since the relation of "congruence modulo m" on the set of integers is reflexive, symmetric and transitive, therefore it is an equivalence relation.

Example 57: *What is the difference between function and relation?*

Let $X = \{1, 2, 3, 4\}$

and $Y = \{1, 5, 9, 11, 15, 16\}$.

Determine which of the following sets are :

(i) relation (ii) function (iii) neither, of X to Y.

(a) $f_1 = \{(x, y) \mid y = x^2, x \in X, y \in Y\}$

(b) $f_2 = \{(1, 1), (2, 11), (3, 1), (4, 15)\}$

(c) $f_3 = \{(1, 5), (2, 9), (3, 1), (4, 5), (2, 11)\}$

(d) $f_4 = \{(1, 1), (2, 7), (3, 5)\}$.

Solution: For difference between function and relation refer

(a) We have $f_1 = \{(1, 1), (3, 9), (4, 16)\}$.

Hence f_1 is a relation of X to Y because $f_1 \subseteq X \times Y$. But f_1 is not a function of X to Y because $2 \in X$ and $2^3 = 4$ which $\notin Y$. Thus under f_1 the image of 2 belonging to X does not exist.

(b) Hence f_2 is a relation of X to Y because $f_2 \subseteq X \times Y$. Also f_2 is a function of X to Y because under f_2 each element of X has a unique image in Y.

(c) Hence, f_3 a relation of X to Y because $f_3 \subseteq X \times Y$. But f_3 is not a function of X to Y because under f_3 the image of $2 \in X$ is not unique. Note the $f_3(2) = 9$ and also $f_3(2) = 11$.

(d) Here f_4 is not a relation of X to Y because $f_4 \not\subseteq X \times Y$. Note that $(2, 7) \in f_4$ but $(2, 7) \notin X \times Y$ because $7 \notin Y$. Also f_4 is not a function of X to Y.

Example 58: *Let N_0 and N_e denote the set of all odd natural numbers and the set of all even natural numbers respectively:*

1. *Is addition a binary operation on N_0?*
2. *Is multiplication a binary operation on N_0?*
3. *Is addition a binary operation on N_e? If yes, find whether this operation has identity element or not.*

Solution:

1. The sum of two odd natural numbers is an even natural number. Thus if $x, y \in N_0$, then $x + y \notin N_0$. Therefore addition is not a binary operation on the set N_0.
2. The product of two odd natural numbers is also an odd natural number and is unique. Thus,

   $x, y \in N_0, \forall x, y \in N_0.$

   Therefore multiplication is a binary operation on the set N_0.
3. The sum of two even natural numbers is an even natural number and is unique. Thus

   $x + y \in N_e, \forall x, y \in N_e$

   Therefore addition is a binary operation on the set N_e.

However, the set N_e does not possess identity element for addition because the number 0 which is identity for ordinary addition of numbers is not an element of the set N_e.

EXERCISES

1. Perhaps the most famous of the logical difficulties referred to in the text is *Russell's paradox*. To explain what this is, we begin by observing that a set can easily have elements which are themselves sets, *e.g.*, {1, {2, 3}, 4}. This raises the possibility that a set might well contain itself as one of its elements. We call such a set an *abnormal* set, and any set which does not contain itself as an element we call a *normal* set. Most sets are normal, and if we suspect that abnormal sets are in some way undesirable, we might try to confine our attention to the set N of all normal sets. Someone is now sure to ask, Is N itself normal or abnormal? It is evidently one or the other, and it cannot be both. Show that if N is normal, then it must be abnormal. Show also that if N is abnormal, then it must be normal. We see in this way that each of our two alternatives is self-contradictory, and it seems to be the assumption that N exists' as a set which has brought us to this impasse.

For further discussion of these matters, we refer the interested reader to Wilder or Frenkel and Bar-Hillel. Russell's own account of the discovery of his paradox can be found in Russell.

2. If A, B are two sets, show that $A - B \neq B - A$.
3. If sets $A = \{(x, y) : x^2 + y^2 = 1\}$ and $B = \{(x, y) : x + y = 1$, then find $A \cap B$.
4. If A and B are any subsets, then prove that $A \cap (B - A) = \phi$.
5. For any two sets A and B, prove that $A' - B' = B - A$.
6. For any two sets A and B, prove that
   $(A - B) \cup (B - A) = (A \cup B) - (A \cap B)$.
7. For any two sets A and B, prove the following:
   (i) $A \cap (A' \cup B) = A \cap B$
   (ii) $A - (A - B) = A \cap B$
   (iii) $A \cap (A \cup B)' = \phi$
   (iv) $A - B = A \Delta (A \cap B)$
8. Shourya is industrious. All the first class students are industrious. Is Shourya first class student?
9. A market research group conducted a survey of 1000 consumers and reported that 720 consumers liked product A and 450 consumer liked product B. What is the least number that must have liked both products?
10. Of the members of three athletic teams in a certain school, 21 are in the basket ball team, 26 in hockey team and 29 in the football team, 14 play hockey and basket ball, 15 play hockey and football, 12 play foot ball and basket ball and 8 play all the three games. How many members are there in all?
11. Prove that the function $f : N \to N$, defined by $f(x) = x^2 + x + 1$ is one-one but not onto.
12. Let $f = \{(3, 1), (9, 3) (12, 4)\}$ and $g = \{1, 3), (3, 3), (4, 9) (5, 9)\}$. Show that fog and gof are both defined. Also, find fog and gof.
13. Let $A = \{a, b, c\}$ $B = \{u, v, w\}$ and let f and g be two functions from A to B and from B to A respectively defined as.
    $f = \{(a, v), (b, u), (c, w)\}$,
    $g = \{(u, b), (v, a), (w, c)\}$.

Show that f and g both are bijections and find fog and gof.

14. Find fog(2) and gof(1) when:

    $f : R \to R\ f(x) = x^2 + 8$ and $g : R \to R : g(x) = 3x^3 + 1$

15. Let R^+ be the set of all non-negative real numbers. If $f : R^+ \to R^+$ and $g : R^+ \to R^+$ are defined as $f(x) = x^2$ and $g(x) = +\sqrt{x}$. Find fog and gof. Are they equal functions?

16. Let $A = \{x \in R \mid -1 \le x \le 1\}$ and $f : A \to A$, $g : A \to A$ be two functions defined by $f(x) = x^2$ and $g(x) = \sin(\pi x/2)$: Show that g^{-1} exists but f^{-1} does not exits. Also find g^{-1}.

17. If $f : R \to R$ be defined by $f(x) = x^3 - 3$, then prove that f^{-1} exists and find a formula for f^{-1}. Hence, find $f^{-1}(24)$ and $f^{-1}(5)$.

18. Let A and B be two sets each with finite number of elements. Assume that there is an injective map from A to B and that there is an injective map from B to A. Prove that there is a bijection from A to B.

19. If $f : Q \to Q$, $g : Q \to Q$ are two functions defined by $f(x) = 2x$ and $g(x) = x + 2$. Show that f and g are bijection maps. Verify that $(g\,of)^{-1} = f^{-1}og^{-1}$.

20. If $f : A \to A$, $g : A \to A$ are two bijections then prove that

    (i) fog is an injection

    (ii) fog is a surjection.

21. Find : (i) 200 (mod 20),

    (ii) 10 (mod 3)

    (iii) –347 (mod 6),

    (iv) 555 (mod 11)

22. Find : (i) 3! (3! + 2!),

    (ii) 30!/28!,

    (iii) 6!/5!

23. Let p and q be integers and suppose M (p, q) is defined recursively by

$$M(p, q) = \begin{cases} 5 \text{ if } p < q \\ M(p - q, q + 2) + a \text{ if } p \ge q \end{cases}$$

Find M (2, 7), M (5, 3) and M (15, 2).

24. By definition of the Ackermann function, find A (3, 1) and A (2, 3)
25. For the Ackermann function, prove that A (2, 3) = 2x = 3.
26. Find the value of :
    (i) $\lceil 6 \rceil$,
    (ii) $\lceil -8 \rceil$
    (iii) $\lfloor 8.3 \rfloor$
    (iv) $\lfloor -8.7 \rfloor$
27. Let $f(x) = \lfloor x/2 \rfloor + \lfloor x/3 \rfloor$ for $x \in N$. Calculate f(x) for $0 \le x \le 10$ and for x = 73.
28. Find $\lfloor 5/2 \rfloor$, $\lfloor (5/2)^2 \rfloor$ and $(\lfloor 5/2 \rfloor)^2$.
29. A function $f : Z^+ \to Z$ is defined by

$$f(x) = \begin{cases} 1 & \text{if } x = 1 \\ f(x/2) & \text{if } x \text{ is even} \\ 1 + f(5x - 9), & \text{if } x \text{ is odd and } x > 1 \end{cases}$$

30. Let A = set of vowels in the English alphabets. Evaluate
    (i) $f_A(o)$
    (ii) $f_A(h)$
    (iii) $f_A(e)$
    (iv) $f_A(b)$

    When f_A is the characteristic function of A.
31. Use the definition of O notation prove that
    (i) $7x^3 - 3x^2$ is $O(x^3)$
    (ii) $\lceil x^2 \rceil$ is $O(x^2)$
    (iii) $1^2 + 2^2 + ... + n^2$ is $O(x^3)$.
32. If $f(x) = 4x^3 - 3x^2$ and $g(x) = x^4$. Prove that f and g are of the same order.
33. If $f(x) = 3x^2 + 4x + 1$ and $g(x) = x^2$ for $x \ge 1$.

    Show that $f(x) = \Theta\, g(x)$.
34. Show that the functions f and g defined by $f(n) = 4n^4 - 3n^2$ and $g(n) = n^4$, where n is a positive integer, have the same orders.

35. Define a function as a set of ordered pairs.

36. What is the fundamental difference between a relation and a function? Is every relation a function?

37. Let A = {–2, –1, 0, 1, 2} and $f : A \to Z$ be a function defined by $f(x) = x^2 - 2x - 3$. Find

    (a) range of f , i.e. f(A)

    (b) pre-images of 6, –3 and 5.

38. If a function $f : R \to R$ be defined by

$$f(x) = \begin{cases} 3x - 2, & x < 0, \\ 1, & x = 0, \\ 4x + 1, & x > 0, \end{cases}$$

    Find f(1), f(–1), f(0), f(2)

39. Let $f : R \to R$ and $g : C \to C$ be two functions defined as $f(x) = x^2$ and $g(x) = x^2$. Are they equal functions?

40. Classify the following functions as injection, Surjection or bijection:

    (i) $f : R \to R$, $f(x) = |x|$

    (ii) $f : Z \to Z$, $f(x) = x^2 + x$

    (iii) $f : Z \to Z$, $f(x) = x = 5$

    (iv) $f : R \to R$, $f(x) = \sin x$

    (v) $f : R \to R$, $f(x) = x^3 + 1$

    (vi) $f : R \to R$, $f(x) = x^3 - x$

    (vii) $f : R \to R$, $f(x) = \sin^2 x + \cos^2 x$

    (viii) $f : Q-\{3\} \to Q$, $f(x) = \dfrac{2x + 3}{x - 3}$

    (ix) $f : Q \to Q$, $f(x) = x^3 + 1$

    (x) $f : R \to R$, $f(x) = 5x^3 + 4$

41. Show that the function $f : N \to N$ given by

    $f(n) = n - (-1)^n$ for all $n \in N$ is one-one and onto.

42. Let X = {1, 2, 3, 4, 5, 6, 7, 8, 9} be any set and its three subsets be given by,

    A = {1, 4, 7, 8}, B = {4, 6, 8, 9}, and C = {3, 4, 5, 7}, then find

(i) $A \cap (B - C)$

(ii) $A - (B \cup C)$

(iii) $A' \cap (B' - C')$

(iv) $A \cup (B \cap C)$.

43. If $A = \{a, b, c, d, e\}$, $B = \{a, c, e, g\}$ and $C = \{b, e, f, g\}$, then verify that:

(i) $(A \cup b) \cap C = (A \cap C) \cup (B \cap C)$;

(ii) $A - (B \cup C) = (A - B) \cap (A - C)$;

(ii) $A \cap (B - C) = (A \cap B) - (A \cap C)$;

44. Give an example of three sets A, B and C such that :

$A \cap B \neq \phi$, $B \cap C \neq \phi$, $A \cap C \neq \phi$, and $A \cap B \cap C \neq \phi$.

45. The set X consists of all points within and on the unit circle $x^2 + y^2 = 1$, whereas the set Y consists of all point on and inside the rectangular boundary $x = 0$, $x = 1$, $y = -1$ and $y = 1$. Determine the sets $X \cup Y$ $X \cap Y$. Illustrate your answer by diagrams.

46. For any sets A, B and C prove that:

(i) $A - B = A \cap B'$

(ii) $A - (A \cap B) = (A \cap B')$

(iii) $A \cap (B - A = \phi$

(iv) $(A - B) \cup (B - A) = (A \cup B) - (A \cap B)$

47. For any sets A, B, C prove that:

(i) $A - (B \cup C) = (A - B) \cap (A - C)$

(ii) $A - (B \cap C) = (A - B) \cup (A - C)$

(iii) $A \cap (A' \cup B) = A \cap B$.

48. For any sets A and B prove that:

(i) $A \cup B = A \cap B \Leftrightarrow A = B$

(ii) $A \cup B = \phi \Leftrightarrow A = \phi$ and $B = \phi$

49. Prove that:

(i) $A - B\ \phi \Leftrightarrow A \subseteq B$

(ii) $A' \subseteq B \Leftrightarrow A \cup B = X$, the universal set.

50. Prove that:

(i) $A \supseteq B \Rightarrow (A - C) \supseteq (B - C)$

(ii) $A \subseteq B \Rightarrow A \cup B = B$ and $A \cap B = A$

51. If $A \cup B = A \cup C$ and $A \cap B = A \cap C$, then show that $B = C$.

52. On the real line, if $A =]\ 0, 3\ [$ and $B =]\ 1, 4]$, then find A', B', $A \cup B$, $A \cap B$ and $A - B$.

53. If $A = \{x \in R: | x - 3 | <\}$ and $B = \{x \in R: | x | \leq 2\}$, find $A \cup B$, $A \cap B$, $A - B$ and $B - A$.

54. Prove or disprove the following:

 (i) $P(A) \cap P(B) = P(A \cap B)$

 (ii) $(A \cup B) - C = A \cup (B - C)$

 (iii) $x \in (A \cap B')' \Rightarrow x \in B$

 (iv) $A \cup B = A \cup C \Leftrightarrow B = C$

55. For any three sets A, B and C prove that:

 (i) $A - (B - C) = (A - B) \cup (A \cap C)$

 (ii) $A \cap (B - C) = (A \cap B) - C$.

56. Decide which ones of the three properties of reflexivity, symmetry, and transitivity are true for each of the following relations in the set of all positive integers: $m \leq n$, $m < n$, m divides n. Are any of these equivalence relations?

57. N be the set of all natural numbers. Prove that each one of the relations R_1 and R_2 on $N \times N$, defined by

$$(a, b)\ R_1\ (c, d) \Leftrightarrow a + d = b + c;$$

$$\text{and } (a, b)\ R_2\ (c, d) \Leftrightarrow ad = bc$$

 is (a, b) equivalence relation.

58. The symbol we have used for set inclusion is similar to that used for the familiar order relation on the real line: if x and y are real numbers, $x \leq y$ means that $y - x$ is non-negative. The order relation on the real line has all the properties mentioned in the text:

 (1') $x \leq x$ for every x;

 (2') $x \leq y$ and $y \leq x \Rightarrow x = y$;

 (3') $x \leq y$ and $y \leq s \Rightarrow x \leq z$.

 It also has an important additional property:

 (4') for any x and y, either $x \leq y$ or $y \leq x$. Property (4') says that any two real numbers are comparable with respect to the relation in question, and it leads us to call the order relation on the real line a *total* (or *linear*) *order relation.* Show by an example that this property is not possessed by set inclusion. It is for this reason that set inclusion is called a *partial order relation.*

59. Define a set and give examples to illustrate the difference between a collection and a set. What are different ways to specify a set? Give examples.

60. Find out which of the following sets are null or singleton.

    (i) A = {x : x is a letter before a in alphabets}

    (ii) A = {x : x + 16 = 16}

    (iii) A = {x : x^2 = 4, x is an odd integer}

61. In the set of all natural numbers, prove that the relation R, defined by

    $$a \; R \; b \Leftrightarrow (1 + ab) > 0$$

    is reflexive, and symmetric but not transitive.

62. Check, which of the following relations on the set of all real numbers are equivalence relations. Find in each case, if any property is violated:

    (i) a R b $\Leftrightarrow$ |a| = |b|;

    (ii) a R b $\Leftrightarrow$ a is a multiple of b;

    (iii) a R b $\Leftrightarrow$ $a^2 + b^2 = 1$;

    (iv) a R b $\Leftrightarrow$ |a – b| > 0;

    (v) a R b $\Leftrightarrow$ a/b, *i.e.,* a is a factor of b.

63. Define a set and give examples to illustrate the difference between a collection and a set. What are different ways to specify a set? Give examples.

64. Find out which of the following sets are null or singleton.

    (i) A = {x: x is a letter before a in alphabets}

    (ii) A = {x: x + 16 = 16}

    (iii) A = {x: x^2 = 4, x is an odd integer}

65. Which of the following sets are different from others:

    (i) ϕ,

    (ii) {0}

    (iii) {ϕ}

66. Show that the set of even integers is a subset of the set of integers.
67. $A = \{a, b, c, d, e\}$, $B = \{a, c, e, g\}$

    and $C = \{b, e, f, g\}$, prove that

    $(A \cup B) \cap C = (A \cap C) \cup (B \cap C)$.
68. Define a relation. Give an example of a relation which is
    (i) reflexive but neither symmetric nor transitive;
    (ii) transitive but neither reflexive nor symmetric;
    (iii) symmetric and transitive but not reflexive;
    (iv) symmetric and reflexive but not transitive.
69. Under what conditions would each of the following be true?
    (a) $A \cup B = \phi$,
    (b) $A \cup \phi = \phi$,
    (c) $A \cap U = U$,
    (d) $A \cup \phi = U$,
    (e) $A \cap B = A$,
    (f) $A \cup A = U$,
    (g) $A \cup B = A \cap B$,
    (h) $A \cap B = \phi$,
70. If A and B are any two sets, then prove that $A - B$, $A \cap B$ and $B - A$ are pairwise disjoint.
71. If $A = \{0, 1\}$, $B = \{1, 2, 3,\}$, show that $A \times B = B \times A$.
72. Prove that $(A \cup B) \cap B' = A$, if and only if $A \cap B = \phi$.
73. If $A = \{x: x \text{ is a factor of } 20\}$

    $B = \{x: x \text{ is a multiple of 5 and } x < 20\}$

    Find $A - B$ and $B - A$.
74. If A, B are two sets, show that $A - B \neq B - A$.
75. If sets $A = \{(x, y): x^2 + y^2 = 1\}$ and $B = \{(x, y): x + y = 1$, then find $A \cap B$.
76. For any two sets A and B, show that $A \times B$ and $B \times A$ have one element in common iff A and B have one element in common.
77. If A and B are any subsets, then prove that $A \cap (B - A) = \phi$.

78. For any two sets A and B, prove that $A' - B' = B - A$.

79. If $A \cup B = A \cup C$ and $A \cap B = A \cap C$, then show that $B = C$.

80. Prove that:

(i) $A \supseteq B \Rightarrow (A - C) \supseteq (B - C)$

(ii) $A \subseteq B \Rightarrow A \cup B = B$ and $A \cap B = A$

81. Prove that:

(i) $A - B\ \phi \Leftrightarrow A \subseteq B$

(ii) $A' \subseteq B \Leftrightarrow A \cup B = X$, the universal set.

82. The graph of a mapping $f:X \rightarrow V$ is a subset of the product $X \times Y$. What properties characterize the graphs of mappings among all subsets of X X V?

83. Let X and Y be non-empty sets. If A_1 and A_2 are subsets of X, and B_1 and B_2 subsets of Y, show the following:

$$(A_1 \times B_1) \cap (A_2 \times B_2) = (A_1 \cap A_2) \times (B_1 \cap B_2);$$

$$(A_1 \times B_1) - (A_2 \times B_2) = (A_1 - A_2) \times (B_1 - B_2)$$
$$\cup (A_1 \cap A_2) \times (B_1 - B_2) \cup (A_1 - A_2) \times (B_1 \cap B_2).$$

84. Let X and Y be non-empty sets, and let A and B be rings of subsets of X and Y, respectively. Show that the class of all finite unions of sets of the form $A \times B$ with $A\ \varepsilon\ A$ and $B\ \varepsilon\ B$ is a ring of subsets of $X \times Y$.

85. Under what conditions would each of the following be true?

(a) $A \cup B = \phi$,

(b) $A \cup \phi = \phi$,

(c) $A \cap U = U$,

(d) $A \cup \phi = U$,

(e) $A \cap B = A$,

(f) $A \cup A = U$,

(g) $A \cup B = A \cap B$,

(h) $A \cap B = \phi$.

86. If A = {a, b}, B {2, 3}, C {3, 4},

find (i) $A \times (B \cup C)$, (ii) $A \times (B \cap C)$.

87. Prove that $A \times B = \phi$ iff either $A = \phi$, or $B = \phi$.

88. If A and B are any two sets, then prove that $A - B$, $A \cap B$ and $B - A$ are pairwise disjoint.

89. Prove that $(A \cup B) \cap B' = A$, if and only if $A \cap B = \phi$.

90. If A = {x : x is a factor of 20}

    B = {x : x is a multiple of 5 and x < 20}

    Find $A - B$ and $B - A$.

91. Define cartesian product of two sets. If A, B and C are sets and $A \subset B$, then prove that $A \times C \subset B \times C$.

92. Using mathematical induction, prove the following.

    (i) $1^2 + 2^2 + 3^2 + \ldots\ldots + n^2 = 1/6\, n\,(n + 1)\,(2n + 1)$

    (ii) $1 . 2 + 2 . 3 + \ldots\ldots + n\,(n + 1) = n/3\,(n + 1)\,(n + 2)$

    (iii) $\dfrac{1}{1,3} + \dfrac{1}{3,5} + \dfrac{1}{5,7} + \ldots\ldots + \dfrac{1}{(2n-1)\,(2n+1)} = n/2n + 1$

    (iv) $\dfrac{1}{2} + \dfrac{1}{4} + \dfrac{1}{8} + \ldots\ldots + \dfrac{1}{2^n} = 1 - \dfrac{1}{2^n}$

    (v) $2^n > n,\ n \in N$

    (vi) $2^n > 3^n,\ n \in N$

93. Use mathematical induction to prove the following.

    (i) $n\,(n + 1)\,(n + 2)$ is a multiple of 6.

    (ii) $3^{2n} - 1$ is divisible by 8.

    (iii) $10^n + 3.4^{n+2} + 5$ is divisible by 9

    (iv) $9^n - 8n - 1$ is a multiple of 64.

    (v) for all positive integers n, prove that :

    $\dfrac{n^7}{7} + \dfrac{n^5}{5} + \dfrac{2n^3}{3} - \dfrac{n}{105}$ is an integer.

94. (a) Let U be the single-element set {1}. There are two subsets, the empty set ϕ and {1} itself. If A and B are arbitrary subsets of U, there are four possible relations of the form $A \subseteq B$. Count the number of true relations among these.

    (b) Let U be the set {1, 2}. There are four subsets. List them. If A and B are arbitrary subsets of U, there are 16 possible relations of the form $A \subseteq B$. Count the number of true ones.

(c) Let U be the set $\{1, 2, 3\}$. There are 8 subsets. What are they? There are 64 possible relations of the form $A \subseteq B$. Count the number of true ones.

(d) Let U be the set $\{1, 2, \ldots, n\}$ for an arbitrary positive integer n. How many subsets are there? How many possible relations of the form $A \subseteq B$ are there? Can you make an informed guess as to how many of these are true?

95. Give an example of a relation which is:

(a) reflexive but not symmetric or transitive;

(b) symmetric but not reflexive or transitive;

(c) transitive but not reflexive or symmetric;

(d) reflexive and symmetric but not transitive;

(e) reflexive and transitive but not symmetric;

(f) symmetric and transitive but not reflexive.

96. Let X be a non-empty set and $\sim$ a relation in X. The following purports to be a proof of the statement that if this relation is symmetric and transitive, then it is necessarily reflexive: $x \sim y \Rightarrow y \sim x$; $x \sim y$ and $y \sim x \Rightarrow x \sim x$; therefore $x \sim x$ for every x. In view of Problem 5f, this cannot be a valid proof. What is the flaw in the reasoning?

97. Let X be a non-empty set. A relation $\sim$ in X is called *circular* if $x \sim y$ and $y \sim z \Rightarrow z \sim x$, and *triangular* if $x \sim y$ and $x \sim z \Rightarrow y \sim z$. Prove that a relation in X is an equivalence relation $\Leftrightarrow$ it is reflexive and circular $\Leftrightarrow$ it is reflexive and triangular.

98. Define a relation. When a relation R on a set A is known as symmetric, reflexive, transitive and antisymmetric? Give an example for each.

99. Show that in the set of all real numbers, the relation 'greater than' is transitive but not reflexive.

100. Show that the relation R in the set of natural numbers N defined by a R b if a divides b, is reflexive and transitive but not symmetric'.

101. Give an example of the relation which is.

(a) reflexive, symmetric but not transitive.

(b) symmetric, transitive but not reflexive.

(c) symmetric but neither transitive nor reflexive.

(d) neither symmetric nor reflexive, nor transitive.

102. If R and S are equivalence relations on a set A, prove that $R \cap S$ is an equivalence relation in A.

103. Which of the following relations in the set of real numbers are equivalence relations

    (a) a R b if $|a| = |b|$,

    (b) a R b if $|a| \geq |b|$,

    (c) a R b if $a - b \geq 0$?

104. S is the set of real numbers and a R b, if $a = \pm b$, determine whether R is an equivalence relation?

105. Prove that if a relation R is transitive then its inverse relation R^{-1} is also transitive.

106. Find all the partitions of {a, b, c}.

107. Let $f: X \to Y$ be an arbitrary mapping. Define a relation in X as follows: $x_i \sim x_2$ means that $f(x_1) = f(x_2)$. Show that this is an equivalence relation and describe the equivalence sets.

108. In the set R of all real numbers, let $x \sim y$ mean that $x - y$ is an integer. Show that this is an equivalence relation and describe the equivalence sets.

109. Let I be the set of all integers, and let m be a fixed positive integer. Two integers a and 6 are said to be *congruent* modulo m—symbolized by $a \equiv b \pmod{m}$—if a – 6 is exactly divisible by m, *i.e.*, if a – 6 is an integral multiple of m. Show that this is an equivalence relation, describe the equivalence sets, and state the number of distinct equivalence sets.

2

METRIC SPACE

INTRODUCTION

The branches of mathematics—in geometry as well as analysis—it has been found extremely convenient to have available a notion of distance which is applicable to the elements of abstract sets. A *metric space* is nothing more than a non-empty set equipped with a concept of distance which is suitable for the treatment of convergent sequences in the set and continuous functions defined on the set. Our purpose in this chapter is to develop in a systematic manner the main elementary facts about metric spaces. These facts are important for their own sake, and also for the sake of the motivation they provide for our later work on topological spaces. Classical analysis can be described as that part of mathematics which begins with calculus and, in essentially the same spirit, develops similar subject matter much further in many directions. It is a great nation in the World of Mathematics, with many provinces, a few of which are ordinary and partial differential equations, infinite series (especially Power series and Fourier series), and analytic functions of a complex variable. Each of these has experienced enormous growth over a long history, and each is rich enough in content to merit a lifetime of study. Analysis is primarily concerned with limit processes and continuity, so it is not surprising that mathematicians thinking along these lines soon found themselves studying (and generalizing) two elementary concepts: that of a convergent sequence of real or complex numbers, and that of a continuous function of a real or complex variable.

We remind the reader of the definitions. First, a sequence

$$\{x_n\} = \{x_1, x_2, ..., x_n, ...\}$$

of real numbers is said to be *convergent* if there exists a real number x (called the *limit* of the sequence) such that, given $\varepsilon > 0$, a positive integer n_o can be bound with the property that

$$n \geq n_0 \Rightarrow | x_n - x | < \varepsilon.$$

This condition means that x_n must be "close" to x for all "sufficiently large" n, and it is usually symbolized by

$$x_n \to x \text{ or } \lim x_n = x$$

and expressed by saying that x_n *approaches* x or x_n converges to x. Second, a real function f defined on a non-empty subset X of the real line is said to be *continuous* at x_0 in X if for each $\varepsilon > 0$ there exists $\delta > 0$ such that

$$x \text{ in } X \text{ and } |x - x_0| < \delta \Rightarrow |f(x) - f(x_0)| < \varepsilon,$$

and f is said to be *continuous* if it is continuous at each point of X. When X is an interval, this definition gives precise expression to the intuitive requirement that f have a graph without breaks or gaps.

The corresponding definitions for sequences of complex numbers and complex functions of a complex variable are word for word the same. Our purpose in giving these definitions in detail here is a simple one. We wish to point out explicitly that each is dependent for its meaning on the concept of the absolute value of the difference between two real or complex numbers. We wish to observe also that this absolute value is the *distance between the numbers* when they are regarded as points on the real line or in the complex plane. In the course of its development, classical analysis became so complex and varied that even an expert could find his way around in it only with difficulty. Under these circumstances, some mathematicians became interested in trying to uncover the fundamental principles on which all analysis rests. This movement had associated with it many of the great names in mathematics of the last century: Riemann, Weierstrass, Cantor, Lebesgue, Hilbert, Riesz, and others. It played a large part in the rise to prominence of topology, modern algebra, and the theory of measure and integration; and when these new ideas began to percolate back through classical analysis, the brew which resulted was modern analysis. As modern analysis developed in the hands of its creators, many a major theorem was given a simpler proof in a more general setting, in an effort to lay bare its inner meaning. Much thought was devoted to analyzing the texture of the real and complex number systems, which are the context of analysis. It was hoped—and these hopes were well founded—that analysis could be clarified and simplified, and that stripping away superfluous underbrush would give new emphasis to what really mattered from the point of view of the underlying theory.

THE SPACES OF CONTINUOUS FUNCTIONS

We gave a brief description of the metric space $\mathcal{C}[0, 1]$. The reader will recall that the points of this space are the bounded continuous real functions defined on the closed unit interval [0, 1] and that its metric is defined by d(f, g) = sup |f(x) – g(x)f. We have two aims in this section: to generalize this very important example by considering functions defined on an arbitrary metric space, and to place all function spaces of this type in their proper context by giving the details of the structural pattern, which they all have in common with one another. We begin with the second, and define the algebraic systems which are relevant to our present interests.

Let L be a non-empty set, and assume that each pair of elements x and y in L can be combined by a process called *addition* to yield an element z in L denoted by z = x + y. Assume also that this operation of addition satisfies the following conditions:

(1) x + y = y + x;

(2) x + (y + z) = (x + y) + z;

(3) there exists in L a unique element, denoted by 0 and called the *zero element*, or the *origin*, such that x + 0 = x for every x',

(4) to each element x in L there corresponds a unique element in L, denoted by –x and called the *negative* of x, such that x + (–x) = 0.

We adopt the device of referring to the system of real numbers or to the system of complex numbers as the scalars. We now assume that each scalar a and each element x in L can be combined by a process called *scalar multiplication* to yield an element y in L denoted by y = ax in such a way that

(5) α(x + y) = ax + αy;

(6) (α + β)x = αx + βx;

(7) (αβ) = a(αx);

(8) 1.x = x.

The algebraic system L defined by these operations and axioms is called a *linear space*. Depending on the numbers admitted as scalars (only the real numbers, or all the complex numbers), we distinguish when necessary between *real linear spaces and complex linear spaces*. A linear space is often called a *vector space*, and its elements are spoken of as *vectors*.

We are not concerned here with developing the algebraic theory of linear spaces. Our only interest is in making available some pertinent concepts and terminology which are useful as a background against which to view the metric spaces we wish to study. With this in mind, we mention a few simple facts which are quite easy to prove from the axioms for a linear space: 0 + x = x for every x; x + z = y + z ⇒ x = y (hint: add –z to both sides on the right); $\alpha . 0 = 0$ (hint: $\alpha . 0 + \alpha x = \alpha(0 + x) = \alpha x = 0 + \alpha x$); $0 . x = 0$ (hint: $0 . X + \alpha x = (0 + \alpha)x = \alpha x = 0 + ax$); and $(-l)x = -x$ (hint: $x + (-l)x = 1 . x + (-l)x = (1 + (-l))x = 0 . x = 0$). The reader will notice that in the relation $0 . x = 0$ we have used the symbol 0 with two different meanings: as a scalar on the left and as a vector on the right. Several other meanings will be given to this sinᵍle symbol, but fortunately it is always possible to avoid confusion by attending closely to the context in which it occurs. It is convenient to introduce the operation of *subtraction* by using the symbol $x - y$ as an abbreviation for $x + (-y)$; $x - y$ is called the *difference* between x and y.

A non-empty subset At of a linear space L is called a *linear subspace* of L if $x + y$ is in M whenever x and y are and if ax is in M (for any scalar α) whenever x is. Since M is non-empty, $0 . x = 0$ shows that 0 is in M. Since $-x = (-l)x$, $-x$ is in At whenever x is. It will be seen at once that a linear subspace of a linear space is itself a linear space with respect to the same operations.

A *normal linear space* is a linear space on which there is defined a *norm*, *i.e.*, a function which assigns to each element x in the space a real number $\| x \|$ in such a manner that

(1) $\| x \| \geq 0$, and $\| x \| = 0 \Leftrightarrow x = 0$;

(2) $\| x + y \| \leq \| x \| + \| y \|$;

(3) $\| \alpha x \| = | \alpha | \| x \|$.

In general terms, a normed linear space is simply a linear space in which there is available a satisfactory notion of the distance from an arbitrary element to the origin. From (3) and the fact that $-x = (-l)x$, we obtain $\| -x \| = \| x \|$. A normed linear space is a metric space with respect to the induced metric defined by

$$d(x, y) = \| x - y \|.$$

A *Banach space* is a normed linear space which is complete as a metric space. Any closed linear subspace of a Banach space is itself a Banach space with respect to the same algebraic operations and the same norm.

So much for the technical framework. We now turn to the metric spaces which really concern us. They are all *function spaces*, in the sense that they are linear spaces whose elements are functions defined on some non-empty set X with addition and scalar multiplication defined pointwise, *i.e.*, by $(f + g)(x) = f(x) + g(x)$ and $(\alpha f)\{x\} = \alpha f(x)$. We note that the zero element in such a linear space is the constant function 0 whose only value is the scalar 0 and that $(-f)(x) = -f(.x)$.

Suppose, then, that X is an arbitrary non-empty set, and consider the set L of all real functions defined on X. It is clear that L is a real linear space with respect to the operations described above. We now restrict ourselves to the subset B consisting of the bounded functions in L. B is obviously a linear subspace of L, so it is a linear space in its own right. Even more, if we define a norm on B by $\|f\| = \sup |f(x)|$, then B is a Banach space.

We next assume that the underlying set X is a metric space. This enables us to consider the possible continuity of functions defined on X. We define $\mathcal{C}(X, R)$ to be that subset of B which consists of continuous functions. $\mathcal{C}(X, R\}$ is thus the set of all bounded continuous real functions defined on the metric space X.

Theorem 1:

The set $\mathcal{C}(X. \mathcal{R})$ of all bounded continuous real functions defined on a metric space X is a real Banach space with respect to pointwise addition and scalar multiplication and the norm defined by $\|f\| = \sup |f(x)|$.

It is desirable at this stage to make a clear distinction between two types of convergence for sequences of functions. Let X be a metric space, and let $\{f_n\}$ be a sequence of real functions defined on X. If for each x in X it happens that $\{f_n(x)\}$ is a Cauchy sequence of real numbers, then by the completeness of the real number system we can define a limit function f by $f(x) = \lim f_n(x)$. We then say that f_n converges pointwise to f, or that f is the pointwise limit of f_n. It is often important to know what properties of the functions f_n carry over to the limit function f, but unless we strengthen the mode of convergence, very little can be said along these lines. The stronger type of convergence normally needed to conclude anything of interest is called *uniform convergence*. In order to explain what this is, we inspect a little more closely what is involved in pointwise convergence. To say that f_n converges pointwise to f is to say the following: for each point x in X, if $\in > 0$ is given, then a positive integer n_0 can be found such that $|fn(x) - f(x)| < \in$ for all $n \geq no$. In general, the integer no may depend on x as well as $\in$. If, however, for each given $\in$ an integer n_0 can be found which serves for all points x, then we say that f_n *converges uniformly* to f, or that f is the uniform limit of f_n. The reader will observe that these concepts are quite independent of the assumption that X is a metric space and that they are meaningful for functions defined on an arbitrary non-empty set.

It will be seen at once that convergence in the function space $\mathcal{C}(X, R)$ is precisely uniform convergence as we have just defined it. The fact that $\mathcal{C}(X, R)$ is complete can be restated as follows in the language of uniform convergence: if a bounded real function f defined on X is the uniform limit of a sequence $\{f_n\}$ of bounded *continuous* real functions defined on X, then f is also continuous. In other words, in the presence of uniform convergence, continuity carries over from the f_n's to the limit function f.

A moment's thought will convince the reader that the entire discussion given above, beginning with our definition of the linear space L, could perfectly well have been based on complex functions. Without going again through all the details, we state the following theorem and consider it proved.

Theorem 2:

The set $\mathcal{C}(X, C)$ of all bounded continuous complex functions defined on a metric space X is a complex Banach space with respect to pointwise addition and scalar multiplication and the norm defined by

$$\|f\| = \sup |f(x)|.$$

In summary, we associate with each metric space X two linear spaces of continuous functions defined on X. The first—$\mathcal{C}(X, R)$—contains only real functions, and the second—$\mathcal{C}(X, C)$—consists of complex functions. Further, all functions considered are assumed to be bounded, so that the norm defined by $\|f\| = \sup |f(x)|$ is always a real number. In the special case in which X is a closed interval [a, b] on the real line, we write $\mathcal{C}(X, R)$ in the simpler form $\mathcal{C}[a, b]$.

DEFINITION AND SOME EXAMPLES

Let X be a non-empty set. A *metric* on X is a real function d of ordered pairs of elements of X which satisfies the following three conditions:

(1) $d(x, y) \geq 0$, and $d(x, y) = 0 \Leftrightarrow x = y$;

(2) $d(x, y) = d(y, x)$ (*symmetry*);

(3) $d(x, y) \leq d(x, z) + d(z, y)$ (the *triangle inequality*).

The function d assigns to each pair (x, y) of elements of X a non-negative real number d(x, y), which by symmetry does not depend on the order of the elements; d(x, y) is called the distance between x and y. A *metric space* consists of two objects: a non-empty set X and a metric d on X. The elements of X are called the *points* of the metric space (X, d). Whenever it can be done without causing confusion, we denote the metric space (X, d) by the symbol X which is used for the underlying set of points. One should always keep in mind, however, that a metric space is not merely a non-empty set: it is a non-empty set together with a metric. It often happens that several different metrics can be defined on a single given non-empty set, and in this case distinct metrics make the set into distinct metric spaces.

There are many different kinds of metric spaces, some of which play very significant roles in geometry and analysis. Our first example is rather trivial, but it is often useful in showing that certain statements we might wish to make are not true. It also shows that every non- empty set can be regarded as a metric space.

Example 1:

Let X be an arbitrary non-empty set, and define d by

$$d(x,y) = \begin{cases} 0 & \text{if } x = y, \\ 1 & \text{if } x \neq y. \end{cases}$$

The reader can easily see for himself that this definition yields a metric on X.

Our next two examples are the fundamental number systems of mathematics.

Example 2:

Consider the real line R and the real function $| x |$ defined on R. Three elementary properties of this absolute value function are important for our purposes:

(i) $| x | \geq 0$, and $| x | = 0 \Leftrightarrow x = 0$;

(ii) $| -x | = | x |$;

(iii) $| x + y | \leq | x | = | y |$.

We now define a metric on R by

$$d(x, y) = | x - y |.$$

This is called the *usual metric* on R, and the real line, as a metric space, is always understood to have this as its metric. The fact that d actually is a metric follows from the three properties stated above. This is a piece of reasoning which occurs frequently in our work, so we give the details. By

(i) $d(x, y) = | x - y |$ is a non-negative real number which equals $0 \Leftrightarrow x - y = 0 \Leftrightarrow x = y$. By

(ii) $d(x, y) = | x - y | = | -(y - x) | = | y - x | = d(y, x)$. And by

(iii) $d(x, y) = | x - y | = | (x - y) + (z - y) | \leq | x - z | + | z - y | = d(x, z) + d(z, y)$.

Example 3:

The set of points in the preceding example—that is, the set of all bounded continuous real functions defined on the closed unit interval—has another metric which is far more important for our purposes. It is defined by means of

$$\| f \| = \sup \{|f(x)| : x \varepsilon [0, 1]),$$

which we usually write more briefly as

$$\| f \| = \sup |f(x)|,$$

and $d(f, g) = \|f - g\| = \sup |f(x) - g(x)|$.

Properties (i) and (ii) of the norm are obvious, and we ask the reader to prove (iii) in a slightly more general form. This example is typical of a large class of metric spaces which will play a major role in all our work throughout the rest of this book. We denote this space by [0, 1].

So much for the present for specific examples. We now turn to several fundamental principles relating to metric spaces in general.

Let X be a metric space with metric d. Let Y be an arbitrary non- empty subset of X. If the function d is considered to be defined only for points in Y, then (Y, d) is evidently itself a metric space. Y, with d restricted in this way, is called a *subspace* of X. This technique of forming subspaces of a given metric space enables us to obtain an infinity of further examples from the handful described above. For instance, the closed unit interval [0, 1] is a subspace of the real line, as is the set consisting of all the rational points; and the unit circle, the closed unit disc, and the open unit disc are subspaces of the complex plane. Also, the real line itself is a subspace of the complex plane.

It is desirable at this stage to introduce the extended real *number system*, by which we mean the ordinary real number system R with the symbols

$$-\infty \text{ and } +\infty$$

adjoined. An *extended real number* is thus a real number or one of these symbols. We say (by definition) that

$$-\infty < +\infty;$$

also, if x is any real number, then

$$-\infty < x < +\infty.$$

The symbols $-\infty$ and $+ \infty$ add nothing to our understanding of the real numbers. They are used mainly as a notational convenience.

Let A be a non-empty set of real numbers which has an upper bound. We defined what is meant by the least upper bound (or supremum) of A: sup A is the smallest upper bound of A, that is, it is the smallest real number y such that $a \leq y$ for every a in A. With the stated assumptions about A, sup A always exists and is a real number. If A is a non-empty set of real numbers which has no upper bound, and therefore no least upper bound in R, we express this by writing

$$\sup A = +\infty;$$

and if A is the empty subset of R, we put

$$\sup A = -\infty$$

The greatest lower bound (or infimum) of A is defined similarly: if A is non-empty and has a lower bound, inf A is the largest real number x such that $x \leq a$ for every a in A; if A is non-empty and has no lower bound, we put

$$\inf A = -\infty;$$

and if A is empty, we put

$$\inf A = +\infty$$

These remarks illustrate one advantage of the extended real number system: it enables us to speak of sup A and inf A for subsets A of the real line without any restrictions whatever on the nature of A.

Another advantage of having available the symbols $-\infty$ and $+\infty$ is that they make convenient a reasonable extension of our concept of an interval on the real line. The definitions whose scope we are now widening. Let a and b be any two real numbers such that $a \leq b$; then the *closed interval* from a to b is the subset of the real line R defined by

$$[a, b] = \{x : a \leq x \leq b\}.$$

This extends our previous notion in that a closed interval may now consist of a single point (if a = b). If b is a real number and a is an extended real number such that $a \leq b$, then the *open-closed interval* from a to & is

$$(a, b) = \{x : a < x \leq b\}.$$

This allows open-closed intervals of the form $(-\infty, b]$. If a is a real number and b is an extended real number such that $a < b$, then the *closed-open interval* from a to b is

$$[a, b) = \{x : a \leq x < b\}.$$

This permits $[a, +\infty)$ to be considered a closed-open interval. If a and b are extended real numbers such that $a < b$, then the *open interval* from a to b is

$$(a, b) = \{x : a < x < b\}.$$

This adds to the previously defined open intervals those of the form $(-\infty, b)$ where b is real, $(a, +\infty)$ where a is real, and $(-\infty, +\infty)$. Throughout the rest of this book, the term *interval* will always signify one of the four types defined in this paragraph. The extended real numbers a and h are called the *end-points* of these intervals.

We have used the symbols $-\infty$ and $+\infty$ with considerable freedom, and it therefore seems desirable to emphasize that an interval in our present sense is always a non-empty subset of the real number system: it never actually contains either of these symbols.

The very definition of a metric space presents us with the concept of the distance from one point to another. We now define the distance from a point to a set and the diameter of a set.

Let X be a metric space with metric d, and let A be a subset of X. If x is a point of X, then the *distance from* x to A is defined by

$$d(x, A) = \inf [d(x, a) : a \varepsilon A\};$$

that is, it is the greatest lower bound of the distances from x to the points of A. The *diameter* of the set A is defined by

$$d(A) = \sup \{d(a_1, a_2) : a_1 \text{ and } a_2 \varepsilon A\}.$$

The diameter of A is thus the least upper bound of the distances between pairs of its points. A is said to have *finite diameter or infinite diameter* according as d(A) is a real number or $\pm\infty$. We observe that the empty set has infinite diameter, since $d(\phi) = -\infty$. A *bounded set* is one whose diameter is finite. A mapping of a non-empty set into a metric space is called a *bounded mapping* if its range is a bounded set. Several of the simpler facts about these concepts are brought out in the following problems.

CONVERGENCE, COMPLETENESS AND BAIRE'S THEOREM

As we emphasized in the introduction to this chapter, one of our main aims in considering metric spaces is to study convergent sequences in a context more general than that of classical analysis. The fruits of this study are many, and among them is the added insight gained into ordinary convergence as it is used in analysis.

Let X be a metric space with metric d, and let

$$\{x_n\} = \{x_1, x_2, \dots, x_n, \dots\}$$

be a sequence of points in X. We say that $\{x_n\}$ is convergent if there exists a point x m X such that either

(1) for each $\varepsilon > 0$, there exists a positive integer n_0 such that $n \geq n_0 \Rightarrow d(x_n, x) < \varepsilon$; or equivalently,

(2) for each open sphere $S_\epsilon(x)$ centered on x, there exists a positive integer no such that x_n is in $S_\epsilon(x)$ for all $n \geq n_o$.

The reader should observe that the first condition is a direct generalization of convergence for sequences of numbers as defined in the introduction, and that the second can be thought of as saying that each open sphere centered on x contains all points of the sequence from some place on. If we rely on our knowledge of what is meant by a convergent sequence of real numbers, the statement that $\{x_n\}$ is convergent can equally well be defined as follows: there exists a point x in X such that $d(x_n,x) \to 0$. We usually symbolize this by writing

$$x_n \to x,$$

and we express it verbally by saying that x_n approaches x, or that x_n *converges to x*. It is easily seen from condition (2) that the point x in this discussion is unique, that is, that $x_n \to y$ with $y \neq x$ is impossible. The point x is called the *limit* of the sequence $\{x_n\}$, and we sometimes write $x_n \to x$ in the form

$$\lim x_n = x.$$

The statements $x_n \to x$ and $\lim x_n = x$

mean exactly the same thing, namely, that $(x_n\}$ is a convergent sequence with limit x.

Every convergent sequence $\{x_n\}$ has the following property: for each $\varepsilon > 0$, there exists a positive integer n_0 such that $m, n \geq n_0 \Rightarrow d(x_m, x_n) < \varepsilon$. For if $x_n \to x$, then there exists a positive integer no such that $n_o \geq n_0$ $\Rightarrow d(x_n, x) < \varepsilon/2$, and from this we see that

$$m,n \geq n_0 \Rightarrow d(x_m, x_n) \leq d(x_m, x) + d(x, x_n) < \varepsilon/2 + \varepsilon/2 = \epsilon.$$

A sequence with this property is called a *Cauchy sequence*, and we have just shown that every convergent sequence is a Cauchy sequence. Loosely speaking, this amounts to the statement that if the terms of a sequence approach a limit, then they get close to one another. It is of basic importance to understand that the converse of this need not be true, that is, that a Cauchy sequence is not necessarily convergent. As an example, consider the subspace X = (0, 1] of the real line. The sequence defined by $x_n = 1/n$ is easily seen to be a Cauchy sequence in this space, but it is not convergent, since the point 0 (which it wants to converge to) is not a point of the space. The difficulty which arises in this example stems from the fact that the notion of a convergent sequence is not intrinsic to the sequence itself, but also depends on the structure of the space in which it lies. A convergent sequence is not convergent "on its own"; it must converge to some point in the space. Some writers emphasize the distinction between convergent sequences and Cauchy sequences by calling the latter "intrinsically convergent" sequences.

A *complete metric* space is a metric space in which every Cauchy sequence is convergent. In rough terms, a metric space is complete if every sequence in it which tries to converge is successful, in the sense that it finds a point in the space to converge to. The space (0, 1] mentioned above is not complete, but it evidently can be made so by adjoining the point 0 to it to form the slightly larger space [0, 1]. As a matter of fact, any metric space, if it isn't already complete, can be made so by suitably adjoining additional points.

It is a fundamental fact of elementary analysis that the real line is a complete metric space. The complex plane is also complete, as we see from the following argument. Let $\{z_n\}$, where $z_n = a_n + ib_n$ be a Cauchy sequence of complex numbers. Then $\{a_n\}$ and $\{b_n\}$ are them- selves Cauchy sequences of real numbers, since,

$$\text{and } |a_m - a_n| \leq |z_m - z_n|$$

$$|b_m - b_n| \leq |z_m - z_n|$$

By the completeness of the real line, there exist real numbers a and b such that $a_n \to a$ and $b_n \to b$. If we now put $z = a + ib$, then we see that $z_n \to z$ by means of

$$\begin{aligned} |z_n - z| &= |(a_n + ib_n) - (a + ib)| \\ &= |(a_n - a) + i(b_n - b)| \\ &\leq |a_n - a| + |b_n - b| \end{aligned}$$

and the fact that both final terms on the right approach 0. The completeness of the complex plane thus depends directly on the completeness of the real line. In this space a Cauchy sequence must be constant (*i.e.*, it must consist of a single point repeated) from some place on, and it converges with that point as its limit.

The problem of completing this space leads to the modern theory of Lebesgue integration, and it would carry us too far afield to pursue this matter to its natural conclusion.

On the other hand, the space C[0, 1] defined. The completeness of this space, and of others similar to it, is one of the major focal points of topology and modern analysis.

The terms *limit* and *limit point* are often a source of confusion for people not thoroughly accustomed to them. On the real line, for instance, the constant sequence (1, 1, ... , 1, ...) is convergent with limit 1; but the set of points of this sequence is the set consisting of the single element 1, the point 1 is not a limit point of this set. The essence of the matter is that a sequence of points in a set is not a subset of the set: it is a function defined on the positive integers with values in the set, and is usually specified by listing its values, as in $\{x_n\} = [x_1, x_2, \dots, x_n, \dots\}$, where Xn is the value of the function at the integer n. A sequence may have a limit, but cannot have a limit point; and the set of points of a sequence may have a limit point, but cannot have a limit. The following theorem relates these concepts to one another and is a useful tool for some of our later work.

Theorem 1:

If a convergent sequence in a metric space has infinitely many distinct points, then its limit is a limit point of the set of points of the sequence.

Proof:

Let X be a metric space, and let $\{x_n\}$ be a convergent sequence in X with limit x. We assume that x is not a limit point of the set of points of the sequence, and we show that it follows from this that the sequence has only finitely many distinct points. Our assumption implies that there exists an open sphere $S_r(x)$ centered on x which contains no point of the sequence different from x. However, since x is the limit of the sequence, all x_n's from some place on must lie in $S_r\{x)$, hence must coincide with x. From this we see that there are only finitely many distinct points in the sequence.

Our next theorem guarantees the completeness of many metric spaces which arise as subspaces of complete metric spaces.

Theorem 2:

Let X be a complete metric space, and let Y be a subspace of X. Then Y is complete $\Leftrightarrow$ it is closed.

Proof:

We assume, first that Y is complete as a subspace of X, and we show that it is closed. Let y be a limit point of Y. For each positive integer n, $S_{1/n}(y)$ contains a point y_n in Y. It is clear that $\{y_n\}$ converges to y in X and is a Cauchy sequence in Y, and since Y is complete, y is in Y. Y is therefore closed.

We now assume that Y is closed, and we show that it is complete. Let $\{y_n\}$ be a Cauchy sequence in y. It is also a Cauchy sequence in X, and since X is complete, $\{y_n\}$ converges to a point x in X. We show that x is in Y. If $\{y_n\}$ has only finitely many distinct points, then x is that point infinitely repeated and is thus in Y. On the other hand, if {yn) has infinitely many distinct points, then, by Theorem A, x is a limit point of the set of points of the sequence; it is therefore also a limit point of Y, and since Y is closed, x is in Y.

A sequence $\{A_n\}$ of subsets of a metric space is called a decreasing sequence if

$$A_1 \supseteq A_1 \supseteq A_1 \supseteq \ldots.$$

The following theorem gives conditions under which the intersection of such a sequence is non-empty.

Theorem 3:

(Cantor's Intersection Theorem): *Let X be a complete metric space, and let $\{F_n\}$ be a decreasing sequence of non-empty closed subsets of X such that $d(F_n) \to 0$. Then $F = \bigcap_{n=1}^{\infty} F_n$ Fn contains exactly one point.*

Proof:

It is first of all evident from the assumption $d(F_n) \to 0$ that F cannot contain more than one point, so it suffices to show that F is nonempty. Let x_n be a point in F_n. Since $d(F_n) \to 0$, $\{x_n\}$ is a Cauchy sequence. Since X is complete, $\{x_n\}$ has a limit x. We show that x is in F, and for this it suffices to show that x is in F_n, for a fixed but arbitrary no. If $\{x_n\}$ has only finitely many distinct points, then x is that point infinitely repeated, and is therefore in F_{n0}. If $\{x_n\}$ has infinitely many distinct points, then a: is a limit point of the set of points of the sequence, it is a limit point of the subset $\{x_n : n \geq n_0\}$ of the set of points of the sequence, it is a limit point of F_{n0} and thus (since F_{n0} is closed) it is in F_{n0}.

A subset A of a metric space is said to be *nowhere dense* if its closure has empty interior. It is easy to see that A is nowhere dense $\Leftrightarrow \overline{A}$ does not contain any non-empty open set $\Leftrightarrow$ each non-empty open set has a non-empty open subset disjoint from $\overline{A} \Leftrightarrow$ each non-empty open set has a non-empty open subset disjoint from A $\Leftrightarrow$ each non-empty open set contains an open sphere disjoint from A. If a nowhere dense set is thought of as a set which doesn't cover very much of the space, then our next theorem says that a complete metric space cannot be covered by any sequence of such sets.

Theorem 4:

If $\{A_n\}$ is a sequence of nowhere dense sets in a complete metric space X, then there exists a point in X which is not in any of the A_n's.

Proof:

For the duration of this proof, we abandon our usual notations for open spheres and closed spheres. Since X is open and A_1 is nowhere dense, *there is an open sphere S_1 of radius less than 1 which is disjoint from A_1. Let F_1 be the concentric closed sphere whose radius is one-half that of S_1, and consider its interior. Since A_2 is nowhere dense, Int (F_1) contains an open sphere S_2 of radius less than 1/2 which is disjoint from As. Let F_2 be the concentric closed sphere whose radius is one-half that of S_2, and consider its interior. Since A_3 is nowhere dense, Int (F_2) contains an open sphere S_3 of radius less than 1/4 which is disjoint from A_3. Let F_3 be the concentric closed sphere whose radius is one-half that of S_3. Continuing in this way, we get a decreasing sequence $\{F_n\}$ of non-empty closed subsets of X such that $d(F_n) \to 0$. Since X is complete, Theorem C guarantees that there exists a point x in X which is in all the F_n's. This point is clearly in all the S_n's, and therefore (since S_n is disjoint from A_n) it is not in any of the A_n's.*

For our purposes, the following equivalent form of Theorem D is often more convenient.

Theorem 5:

If a complete metric space is the union of a sequence of its subsets, then the closure of at least one set in the sequence must have non- empty interior.

Theorems D and E are really one theorem expressed in two different ways. We refer to both (or either) as *Baire's theorem*. This theorem is admittedly rather technical in nature.

He will find, however, that a need for it crops up from time to time, and when this need arises, Baire's theorem is an indispensable tool.

CONTINUOUS MAPPINGS

In the previous section we extended the idea of convergence to the context of a general metric space. We now do the same for continuity.

Let X and Y be metric spaces with metrics d_1 and d_2, and let f be a mapping of X into Y. f is said to be *continuous at a point* x_0 in X if either of the following equivalent conditions is satisfied:

(1) for each $\in > 0$ there exists $\delta > 0$ such that $d_1(x, x_0) < \delta \Rightarrow d_2 f(x), f(x_0)) < \in$;

(2) for each open sphere $S_\in(f(x_0))$ centered on $f(x_0)$ there exists an open sphere $S_\delta(x_0)$ centered on a_0 such that $f(S_\delta(x_0)) \subseteq S_\in(f(x_0))$.

The reader will notice that the first condition generalizes the elementary definition given in the introduction to this chapter, and that the second translates the first into the language of open spheres.

Our first theorem expresses continuity at a point in terms of sequences which converge to the point.

Theorem 1:

Let X and Y be metric spaces and f a mapping of X into Y. Then f is continuous if and only if $x_n \to x \Rightarrow f(x_n) \to f(x)$.

This result shows that continuous mappings of one metric space into another are precisely those which send convergent sequences into convergent sequences, or, in other words, which *preserve convergence.* Our next theorem characterizes continuous mappings in terms of open sets.

Theorem 2:

Let X and Y be metric spaces and f a mapping of X into Y. Then f is continuous $\Leftrightarrow f^{-1}(G)$ *is open in X whenever G is open in Y.*

Proof:

We first assume that f is continuous. If G is an open set in Y, we must show that $f^{-1}(G)$ is open in X. $f^{-1}(G)$ is open if it is empty, so we may assume that it is non-empty. Let x be a point in $f^{-1}(G)$. Then f(x) is in G, and since G is open, there exists an open sphere $S_\epsilon(f(x))$ centered on f(x) and contained in G. By the definition of continuity, there exists an open sphere $S_\delta(x)$ such that $f(S_\delta(x)) \subset S_\epsilon(f(x))$. Since $S_\epsilon(f(x)) \subseteq G$, we also have $f(S_\delta(x)) \subseteq G$, and from this we see that $S_\delta(x) \subseteq f^{-1}(G)$. $S_\delta(x)$ is therefore an open sphere centered on x and contained in $f^{-1}(G)$, so $f^{-1}(G)$ is open.

We now assume that $f^{-1}(G)$ is open whenever G is, and we show that f is continuous. We show that f is continuous at an arbitrary point x in X. Let $S_\epsilon(f(x))$ be an open sphere centered on f(x). This open sphere is an open set, so its inverse image is an open set which contains x. By this, there exists an open sphere $S_\delta(x)$ which is contained in this inverse image. It is clear that $f(S_\delta(x))$ is contained in $S_\epsilon(f(x))$, so f is continuous at x. Finally, since x was taken to be an arbitrary point in X, f is continuous.

The fact just established—that continuous mappings are precisely those which pull open sets back to open sets—will be of great importance for all our work.

We now come to the useful concept of *uniform continuity*. In order to explain what this is, we examine the definition of continuity expressed in condition (1) at the beginning of this section. Let X and Y be metric spaces with metrics d_1 and d_2, and let f be a mapping of X into Y. We assume that f is continuous, that is, that for each point x_0 in X the following is true: given

$\in > 0$, a number $\delta > 0$ can be found such that $d_1(x, x_0) < \delta \Rightarrow d_2(f(x), f(x_0)) < \in$. The reader is no doubt familiar with the idea that if xa is held fixed and e is made smaller, then, in general, δ has to be made correspondingly smaller. Thus, in the case of the real function f defined by $f(x) = 2x$, δ can always be chosen as any positive number $\leq \in/2$, and no larger δ will do. In general, therefore, δ depends on $\in$. Let us return to our examination of the definition. It says that for our given $\in$, a δ can be found which "works" in the above sense at the particular point x_0 under consideration.

But if we hold $\in$ fixed and move to another point x_0, then it may happen that this δ no longer works; that is, it may be necessary to take a smaller δ to satisfy the requirement of the definition. We see in this way that δ nay well depend, in general, not only on $\in$ but also on x_0. Uniform continuity is essentially continuity plus the added condition that for each $\in$ we can find a δ which works *uniformly over the entire space* X, in the sense that it does not depend on x_0.

The formal definition is as follows. If X and Y are metric spaces with metrics d_1 and d_2, then a mapping f of X into Y is said to be *uniformly continuous* if for each $\in > 0$ there exists $\delta > 0$ such that $d_1(x, x') < \delta \Rightarrow d_2(f(x), f(x')) < \in$. It is clear that any uniformly continuous mapping is automatically continuous. The reader will observe that the above real function f defined on the entire real line R by $f(x) = 2x$ is uniformly continuous. On the other hand, the function g defined on R by $g(x) = x^2$ is continuous but not uniformly continuous. Similarly, the continuous function h defined on (0, 1) by $h(x) = 1/x$ is not uniformly continuous.

Uniformly continuous mappings—as opposed to those which are merely continuous—are of particular significance in analysis. The following theorem expresses a property of these mappings which is often useful.

Theorem 3:

Let X be a metric space, let Y be a complete metric space, and let A be a dense subspace of X. If is a uniformly continuous mapping of A into Y, then f can be extended uniquely to a uniformly continuous mapping g of X into Y.

Proof:

Let d_1 and d_2 be the metrics on X and Y. If $A = X$, the conclusion is obvious. We therefore assume that $A \neq X$. We begin by showing how to define the mapping g. If x is a point in A, we define g(x) to be f(x). Now let x be a point in $X - A$. Since A is dense, x is the limit of a convergent sequence $\{a_n\}$ in A. Since $\{a_n\}$ is a Cauchy sequence and f is uniformly continuous,

$\{f(a_n)\}$ is a Cauchy sequence in Y. Since Y is complete, there exists a point in Y—we call this point g(x)—such that $f(a_n) \to g(x)$. We must make sure that g(x) depends only on x, and not on the sequence $\{a_n\}$. Let $\{b_n\}$ be another sequence in A such that $b_n \to x$. Then $d_1(a_n, b_n) \to 0$, and by the uniform continuity of f, $d_2(f(a_n), f(b_n)) \to 0$. It readily follows from this that $f(b_n) \to g(x)$.

We next show that g is uniformly continuous. Let $\in > 0$ be given, and use the uniform continuity of f to find $\delta > 0$ such that for a and a' in A we have $d_1(a, a') < \delta \Rightarrow d_2(f(a), f(a')) < \in$. Let x and x' be any points in X such that $d_1(x, x') < \delta$. It suffices to show that $d_2(g(x), g(x')) \leq \in$. Let $\{a_n\}$ and $\{a'_n\}$ be sequences in A such that $a_n \to x$ and $a'_n \to x'$.

By the triangle inequality, we see that $d_1(a_n, a'_n) < d_1(a_n, x) + d_1(x, x') + d_1(x', a'_n)$. This inequality, together with the facts that $d_1(a_n, x) \to 0$, $d_1(x, x') < \delta$, and $d_1(x', a'_n) \to 0$, implies that $d_1(a_n, a'_n) < \delta$ for all sufficiently large n. It now follows that $d_2(f(a_n), f(a'_n)) < \in$ for all sufficiently large n.

$$d_2(g(x), g(x')) = \lim d_2 f(a_n), f(a'_n)),$$

and from this and the previous statement we see that $d_2(g(x), g(x')) \leq \in$.

All that remains is to show that g is unique, and this is easily proved.

There is an important type of uniformly continuous mapping which often arises in practice. If X and Y are metric spaces with metrics di and dt, a mapping f of X onto Y is called an *isometry* (or an *isometric mapping*) if $d_1(x, x') = d_2(f(x), f(x'))$ for all points x and x' in X; and if such a mapping exists, we say that X is *isometric* to Y. It is clear that an isometry is necessarily one-to-one. If X is isometric to Y, then the points of these spaces can be put into one-to-one correspondence in such a way that the distances between pairs of corresponding points are the same. The spaces therefore differ only in the nature of their points, and this is often unimportant. We usually consider isometric spaces to be identical with one another. It is often convenient to be able to use this terminology in the case of mappings which are not necessarily onto. If f is a mapping of X into Y which preserves distances in the above sense, then we call f an isometry of X *into* Y, or an isometry of X *onto* the subspace f(X) of Y. In this situation, we often say that Y contains an *isometric image* of X, namely, the subspace f(X).

OPEN SETS

Let X be a metric space with metric d. If xa is a point of X and r is a positive real number, the *open sphere* $S_r(x_0)$ with *center* x_0 and *radius* r is the subset of X defined by

$$S_r(x_0) = \{x : d\{x, x_0) < r\}.$$

An open sphere is always non-empty, for it contains its center. An open sphere with radius 1 contains only its center. $S_r(x_0)$ is often called the open sphere with radius r centered on x_0; intuitively, it consists of all points in X which are "close" to x_0, with the degree of closeness given by r.

A few concrete examples are in order. It should be easy to visualize the open sphere $S_r(x_0)$ on the real line: it is the bounded open interval $(x_0 - r, x_0 + r)$ with mid-point x_0 and total length 2r. Conversely, it is clear that any bounded open interval on the real line is an open sphere, so the open spheres on the real line are precisely the bounded open intervals. The open sphere $S_r(z_0)$ in the complex plane (Fig. 2.1) is the inside of the circle with center z_0 and radius r. Figure 18 illustrates an open sphere in the space $\mathcal{C}[0, 1]$: $S_r(f_0)$ consists of all functions f in $\mathcal{C}[0, 1]$ whose graphs lie within the shaded band of vertical width 2r centered on the graph of f_0. A subset G of the metric space X is called an open set if, given any point x in G, there exists a positive real number r such that $S_r(x) \subseteq G$, that is, if each point of G is the center of some open sphere contained in G. Loosely speaking, a set is open if each of its points is "inside" the set, in the sense made precise by the definition. On the real line, a set consisting of a single point is not open, for each bounded open interval centered on the point contains points not in the set. Similarly, the subset [0, 1) of the real line is not open, because the point 0 in [0, 1) has the property that each bounded open interval centered on it (no matter how small it may be) contains points not in [0, 1), *e.g.*, negative points. If we omit the offending point 0, the resulting bounded open interval (0, 1) is an open set (this is very easy to prove and is a special case of Theorem B below). Further, it is quite clear that any open interval—bounded or not—is an open set, and also that the open intervals are the only intervals which are open sets.

Theorem 1:

In any metric space X, the empty set ϕ and the full space X are open sets.

Proof:

To show that ϕ is open, we must show that each point in ϕ is the center of an open sphere contained in ϕ; but since there are no points in ϕ, this requirement is automatically satisfied. X is clearly open, since every open sphere centered on each of its points is contained in X.

We have seen that [0, 1) is not open as a subset of the real line. However, if we consider [0, 1) as a metric space X in its own right, as a subspace of the real line, then [0, 1) is open as a subset of X, since from this point of view it is the full space. This apparent paradox disappears when we realize that points outside of a given metric space have no relevance to any discussion taking place within the context of that space. A set is open or not open only with respect to a specific metric space containing it, never on its own.

Our next theorem justifies the adjective in the expression "open sphere."

Theorem 2:

Let X be a metric space. A subset G of X is open $\Leftrightarrow$ it is a union of open spheres.

Proof:

We assume first that G is open, and we show that it is a union of open spheres. If G is empty, it is the union of the empty class of open spheres. If G is non-empty, then since it is open, each of its points is the center of an open sphere contained in it, and it is the union of all the open spheres contained in it.

We now assume that G is the union of a class S of open spheres. We must show that G is open. If S is empty, then G is also empty, and by Theorem A, G is open. Suppose that S is non-empty. G is also non-empty. Let a; be a point in G. Since G is the union of the open spheres in S, x belongs to an open sphere $S_r(x_0)$ in S. By Theorem B, x is the center of an open sphere $S_{r_1}(x) \subseteq S_r(x_0)$. Since $S_r(x_0) \subseteq G$, $S_{r_1}(x) \subseteq G$ and we have an open sphere centered on x and contained in G. G is therefore open.

The fundamental properties of the open sets in a metric space are those stated in

Theorem 3:

Let X is a metric space. Then (1) any union of open sets in X is open; and (2) any finite intersection of open sets in X is open.

Proof:

To prove (1), let $(G_1\}$ be an arbitrary class of open sets in X. We must show that $G = \cup_i G_i$ is open. If $\{G_i\}$ is empty, then G is empty, and by Theorem A, G is open. Suppose that $\{G_i\}$ is non-empty. By Theorem C, each G, (being an open set) is a union of open spheres; G is the union of all the open spheres which arise in this way; and by another application of Theorem C, G is open.

To prove (2), let $\{G_i\}$ be a finite class of open sets in X. We must show that $G = \cap_i G_i$ is open. If (G",} is empty, then G = X; and by Theorem A, G is open. Suppose that $\{G_i\}$ is non-empty and that $\{G_i\} = \{G_1, G_2, \ldots G_n\}$

for some positive integer n. If G happens to be empty, then it is open by Theorem A, so we may assume that G is non-empty. Let a; be a point in G. Since x is in each G_i, and each G_i is open, for each i there is a positive real number r, such that $S_{ri}(x) \subseteq G_i$. Let r be the smallest number in the set $\{r_1, r_2, \dots, R_n\}$. This number r is a positive real number such that $S_r(x) \subseteq S_{ri}(x)$ for each i, so $S_r(x) \subseteq G_i$ for each i, and therefore $S_r(x) \subseteq G$. Since $S_r(x)$ is an open sphere centered on x and contained in G, G is open.

The above theorem says that the class of all open sets in a metric space is closed under the formation of arbitrary unions and finite intersections. The reader should clearly understand that Theorem A is an immediate consequence of this statement, since the empty set is the union of the empty class of open sets and the full space is its intersection. The limitation to *finite* intersections in this theorem is essential. To see this, it suffices to consider the following sequence of open intervals on the real line:

$$(-1,1)\left(-\frac{1}{2},\frac{1}{2}\right),\left(-\frac{1}{3},\frac{1}{3}\right)$$

The intersection of these open sets is the set {0} consisting of the single point 0, and this set is not open.

In an arbitrary metric space, the structure of the open sets can be very complicated indeed. Theorem C contains the best information available in the general case: each open set is a union of open spheres. In the case of the real line, however, a description can be given of the open sets which is fairly explicit and reasonably satisfying to the intuition.

Theorem 4:

Every non-empty open set on the real line is the union of a countable disjoint class of open intervals.

Proof:

Let G' be a non-empty open subset of the real line. Let x be a point of G. Since G is open, x is the center of a bounded open interval contained in G. Define I_x to be the union of all the open intervals which contain x and are contained in G. The following three facts are easily proved: I_x is an open interval, which contains x and is contained in G; I_x contains every open interval which contains x and is contained in G; and if y is another point in I_x, then $I_x = I_y$. We next observe that if x and y are any two distinct points of G, then I_x and I_y are either disjoint or identical; for if they have a common point z, then $I_x = I_z$ and $I_y = I_z$, so $I_x = I_y$. Consider the class I of all distinct sets of the form I_x for points x in G. This is a disjoint class of open intervals, and G is obviously its union. It remains to be proved that I is countable.

Let G_r, be the set of rational points in G. G_r is clearly non-empty. We define a mapping f of G_r onto I as follows: for each r in G_r, let f(r) be that unique interval in I which contains r. G_r is countable, and the fact that I is countable.

A firm grasp of the ideas involved in the theory of metric spaces depends on one's capacity to "see" these spaces with the mind's eye. The complex plane is perhaps the best metric space to use as a model from which to absorb this necessary intuitive understanding. When we consider an unspecified set A of complex numbers, we usually imagine it as a region bounded by a curve, as in Fig. 2.3. We think of the point z_1, which is completely surrounded by points of A, as being "inside" the set A, or in its "interior," while z_2 is on the "boundary" of A. More precisely, z_1 is the center of some open sphere contained in A, and each open sphere centered on z_2 intersects both A and its complement A'. We formulate these ideas for a general metric space in the next paragraph and at the end of the next section.

Let X be an arbitrary metric space, and let A be a subset of X. A point in A is called an *interior point* of A if it is the center of some open sphere contained in A; and the *interior* of A, denoted by Int (A), is the set of all its interior points. Symbolically,

$$\text{Int }(A) = \{x : x \in A \text{ and } S_r(x) \subseteq A \text{ for some } r\}.$$

The basic properties of interiors are the following:

(1) Int (A) is an open subset of A which contains every open subset of A (this is often expressed by saying that the interior of A is the largest open subset of A);

(2) A is open $\Leftrightarrow$ A = Int (A);

(3) Int (A) equals the union of all open subsets of A.

CLOSED SETS

Let X be a metric space with metric d. If A is a subset of X, a point x in X is called a *limit point* of A if each open sphere centered on x contains at least one point of A different from x. The essential idea here is that the points of A different from x get "arbitrarily close" to x, or "pile up" at x.

The subset $\left\{1, \frac{1}{2}, \frac{1}{3}, \ldots\right\}$ of the real line has 0 as a limit point; in fact, 0 is its only limit point. The closed-open interval [0, 1) has 0 as a limit point which is in the set and 1 as a limit point which is not in the set; further, every real number x such that $0 < x < 1$ is also a limit point of this set. The set of all integral points on the real line has no limit points at all, whereas every real number is a limit point of the set of all rationals. Every open sphere of radius less than 1 consists only of its center, so no subset of this space has any limit points.

A subset F of the metric space X is called a *closed set* if it contains each of its limit points. In rough terms, a set is closed if its points do not get arbitrarily close to any point outside of it. Among the subsets of the real line mentioned in the preceding paragraph, only the set of integral points is closed. Every subset is closed.

Theorem 1:

In any metric space X, the empty set ϕ and the full space X are closed sets.

Proof:

The empty set has no limit points, so it contains them all and is therefore closed. Since the full space X contains all points, it automatically contains its own limit points and thus is closed.

The following theorem characterizes closed sets in terms of open sets. We already know a good deal about open sets, so that characterization provides us with a useful tool for establishing properties of closed sets.

Theorem 2:

Let X be a metric space. A subset F of X is closed $\Leftrightarrow$ its complement F' is open.

Proof:

Assume first that F is closed. We show that F' is open. If F' is empty, it is open, so we may suppose that F' is non-empty. Let x be a point in F'. Since F is closed and x is not in F, x is not a limit point of F. Since x is not in F and is not a limit point of F, there exists an open sphere $S_r(x)$ which is disjoint from F. $S_r(x)$ is an open sphere centered on x and contained in F', and since x was taken to be any point of F', F' is open.

We now assume that F' is open and show that F is closed. The only way F can fail to be closed is to have a limit point in F'. This cannot happen, for since F' is open, each of its points is the center of an open sphere disjoint from F, and no such point can be a limit point of F.

If x_0 is a point in our metric space X, and r is a non-negative real number, the closed sphere $S_r[x_0]$ with *center* x_0 and *radius* r is the subset of X defined by

$$S_r[x_0] = \{x : d(x, x_0) \leq r\}.$$

$S_r[x_0]$ contains its center, and when r = 0 it contains only its center. The closed spheres on the real line are precisely the closed intervals. In this connection, we observe that though open spheres on the real line are open intervals, there are open intervals which are not open spheres, *e.g.*, $(-\infty, +\infty)$.

The following theorem justifies the adjective in the phrase "closed sphere."

Theorem 3:

In any metric space X, each closed sphere is a closed set.

Proof:

Let $S_r[x_0]$ be a closed sphere in X. By Theorem B, it suffices to show that its complement $S_r[x_0]'$ is open. $S_r[x_0]'$ is open if it is empty, so we may assume that it is non-empty. Let a; be a point in $S_r[x_0]'$. Since $d(x, x_0) > r$, $r_1 = d(x, x_0) - r$ is a positive real number. We take r_1 as the radius of an open sphere Sr_1 (x) centered on x, and we show that $S_r[x_0]'$ is open by showing that $S_{r_1}(x) \subseteq S_r[x_0]'$. Let y be a point in $S_{r_1}(x)$, so that $d(y, x) < r_1$. On the basis of this and the fact that $d(x_0, x) < d(x + y) + d(y, x)$, we see that $d(y, xa_0) > d(x, x_0) - d(y, x) > d(x, x_0) - n = d(x, x_0) - [d(x, x_0) - r] = r$, so that y is in $S_r[x_0]'$.

The main general facts about closed sets are those given in our next theorem.

Theorem 4:

Let X be a metric space. Then (1) any intersection of closed sets in X is closed; and (2) any finite union of closed sets in X is closed.

Proof:

This theorem is an immediate consequence. We prove (1) as follows. If $\{F_i\}$ is an arbitrary class of closed subsets of X and $F = F \cap_i F_i$, then by Theorem B, F is closed if F' is open; but $F' = \cup_i F_i'$ is open, since by Theorem B each F_i' is open. The second statement is proved similarly.

In Theorem E of the previous section, we gave an explicit characterization of the open sets on the real line. We now consider the structure of its closed sets. Among the simplest closed sets on the real line are the closed intervals (which are the closed spheres) and finite unions of closed intervals. Finite sets are included among these, since a set consisting of a single point is a closed interval with equal end-points. What is the character of the most general closed set on the real line? Since closed sets are the complements of open sets, The most general proper closed subset of the real line is obtained by removing a countable disjoint class of open intervals. This process sounds innocent enough, but in fact it leads to some rather curious and complicated examples. One ,of these examples is of particular importance. It was studied by Cantor and is usually called the *Cantor set*. To construct the Cantor set, we proceed as follows (Fig. 2.4). First, denote the closed unit interval [0, 1] by F_1. Next, delete from F_1 the open interval (1/3, 2/3) which is its middle third, and denote the remaining closed set by F_2. Clearly,

$$F_3 = F_2 = \left[0, \frac{1}{3}\right] \cup \left[\frac{2}{3}, 1\right]$$

Next, delete from F_2 the open intervals $\left(\frac{1}{9}, \frac{2}{9}\right)$ and $\left(\frac{7}{9}, \frac{8}{9}\right)$, which are the middle thirds of its two pieces, and denote the remaining closed set by F_3. It is easy to see that,

$$F_3 = \left[0, \frac{1}{9}\right] \cup \left[\frac{2}{9}, \frac{1}{3}\right] \cup \left[\frac{2}{3}, \frac{7}{9}\right] \cup \left[\frac{8}{9}, 1\right]$$

If we continue this process, at each stage deleting the open middle third of each closed interval remaining from the previous stage, we obtain a sequence of closed sets F_n, each of which contains all its successors. The Cantor set F is defined by

$$F = \bigcap_{n=1}^{\infty} F_n,$$

and it is closed by Theorem D. F consists of those points in the closed unit interval [0, 1] which "ultimately remain" after the removal of all the open intervals $\left(\frac{1}{3}, \frac{2}{3}\right), \left(\frac{1}{9}, \frac{2}{9}\right), \left(\frac{7}{9}, \frac{8}{9}\right), \ldots$ What points, do remain? F clearly contains the end-points of the closed intervals which make up each set F_n:

$$0, 1, \frac{1}{3}, \frac{2}{3}, \frac{1}{9}, \frac{2}{9}, \frac{7}{9}, \frac{8}{9} \ldots$$

Does F contain any other points? We leave it to the reader to verify that 1/4 is in F and is not an end-point. Actually, F contains a multitude of points other than the above end-points, for the set of these end-points is clearly countable, while the cardinal number of F itself is c, the cardinal number of the continuum. To prove this, it suffices to exhibit a one-to-one mapping/ of [0,1) into F. We construct such a mapping as follows. Let x be a point in [0,1), and let $x = b_1 b_2 b_3 \ldots$ be its binary expansion. Each b_n is either 0 or 1. Let $t_n = 2b_n$, and regard $t_1 t_2 t_3 \ldots$ as the ternary expansion of a real number f(x) in [0, 1). The reader will easily convince himself that f(x) is in the Cantor set F: since t_1 is 0 or 2, f(x) is not in $\left[\frac{1}{3}, \frac{2}{3}\right)$; since t_2 is 0 or 2, f(x) is not in $\left[\frac{1}{9}, \frac{2}{9}\right)$ or $\left[\frac{7}{9}, \frac{8}{9}\right)$ ete. Also, it is easy to see that the mapping f : [0, 1) → F is onc-to-one. According to this, F contains exactly as many points as the entire closed unit interval [0, 1]. It is interesting to compare this conclusion with the fact that the sum of the lengths of all the open intervals removed is precisely 1, since

$$\frac{1}{3} + \frac{2}{9} + \frac{4}{27} + \ldots = 1$$

It is also interesting to observe (by doing a little arithmetic) that F_{25} is the union of 16,777,216 disjoint closed intervals of the same length which are rather irregularly distributed along [0, 1]. These facts may suffice to indicate that the Cantor set is a very intricate mathematical object and is just the sort of thing mathematicians delight in. We shall encounter this set again from time to time.

We conclude this section by defining two additional concepts which are often useful.

Let X be an arbitrary metric space, and let A be a subset of –X". The *closure* of A, denoted by $\overline{A}$, is the union of A and the set of all its limit points. Intuitively, $\overline{A}$ is A itself together with all other points in X which are arbitrarily close to A. As an example, if $\overline{A}$ is the open unit disc $\{z:|z| < 1\}$ in the complex plane, then A is the closed unit disc $\lfloor z:|s| \leq 1\}$. Thc main facts about closures are the following:

(1) $\overline{A}$ is a closed superset of A which is contained in every closed superset of A (we express this by saying that $\overline{A}$ is the smallest closed superset of A);

(2) A is closed ⇔ A = $\overline{A}$;

(3) A equals the intersection of all closed supersets of A.

Our second concept relates to the discussion of Fig. 2.3 given at the end of the previous section. Again, let X be a metric space and A subset of X. A point in X is called a *boundary point* of A if each open sphere centered on the point intersects both A and A', and the boundary of A is the set of all its boundary points. This concept possesses the following properties:

(1) The boundary of A equals A ∩ A';

(2) The boundary of A is a closed set;

(3) A is closed ⇔ it contains its boundary.

INTERIOR

Let (X, d) be a metric space and $A \subset X$.

A point $X \in A$ is called an interior point of A if $\exists\, r \in R^+$ s.t. $S_{r\,(x)} \subset A$.

The set of all interior points of A is called the interior of A and is denoted by A", or by Int (A).

Thus,

$$A^o = \text{int. } (A) = \{x \in A : S_{r\,(x)} \subset A \text{ for some } r\}.$$

Alternatively, we define

$$A^o = \cup\,(S_{r\,(x)} : S_{r\,(x)} \subset A)$$

Evidently,

(i) A^o is an open set,

For an arbitrary union of open sets is open.

(ii) A^o is the largest open subset of A.

CLOSURE

Let (X, d) be a metric space and $A \subset X$.

The closure of A, denoted by $\overline{A}$ is defined as the intersection of all closed sets that contain A. Symbolically,

$$\overline{A} = \cap\,\{F \subset X : F \text{ is closed, } F \subset A\}. \qquad ...(1)$$

Evidently,

(i) A^o is closed set.

For an arbitrary intersection of closed sets is closed.

(ii) $\overline{A} \subset A$.

(iii) $\overline{A}$ is the smallest closed set which contains A.

Alternatively we define

$$\overline{A} = A \cup D\,(A). \qquad ...(2)$$

A point $x \in \overline{A}$ is called a point of closure of A.

Alternatively, a point $x \in X$ is called a point of closure of A iff $x \in A$ or $x \in D\,(A)$.

Later on we shall show that both the definitions (1) and (2) are equivalent.

DEFINITION OF BOUNDARY OF A POINT

(i) Boundary or Frontier of a set a is denoted by b (A) or F_r (A) and is defined as $b\,(A) = F_r\,(A) = X - A^o \cup (X - A)^o$.

Elements of b (A) are called boundary points of A.

(ii) The exterior of A is defined as the set $(X - A)^o$ and is denoted by ext (A).

Symbolically ext (A) = $X - A^o$.

(iii) A is said to be dense or everywhere dense in X if $\overline{A} = X$.

(iv) A is said to be somewhere dense if $(\overline{A})^o$ [1] f, *i.e.*, if closure of A contains some open set.

(v) A is said to be non where dense (or nonwhere dense set) if $(\overline{A})^o$ = f.

(vi) A metric space (X, d) is said to be separable if \$ A Ì X s.t.

A is countable and $\overline{A} = X$.

(vii) A is said to be dense in itself if $A \subset D(A)$.

UNIFORM CONVERGENCE AND POINTWISE CONVERGENCE

Definition : Let $< f(x_n) >$ be a sequence in a metric space (X, d). This sequence is said to converge pointwise to f (x) for every $x \in X$ that is to say,

$$\lim_{n \to \infty} f_n(x) = f(x) \; \forall \; x \in X.$$

By def. of point,

given $\varepsilon > 0$ and $x \in X$, $\exists \; m \in N$ s.t. $n \geq m$

$$\Rightarrow d(f_n(x), f(x) < \varepsilon$$

Here m = m (ε, x).

If we keep ε fixed and vary x, then we shall get a set of values of m for different $x \in X$. This set of values may or may not have an upper bound. If this set has an upper bound, say n_0, then

$$\forall \; n \geq n_0 \Rightarrow d(f_n(x),$$
$$f(x) < \varepsilon \; \forall \; x \in X.$$

In this case sequence $<f_n(x)>$ is called *uniformly convergent.*

Definition : A sequence defined on a metric space (X,d) is said to be uniformly convergent if given $\varepsilon > 0$, $\exists \; n_0 \in N$

s.t. $n \geq n_0 \Rightarrow d(f_n(x), f(x) < \varepsilon \; \forall \; x \in X$.

Remark : It should be noted that ∃ one m ∀ x in case of uniformly convergent where as in case of ordinary convergence or pointwise convergence, one value of x will correspond to one value of m.

SOME IMPORTANT INEQUALITIES

Lemma 1:

If $1 < p < \infty$ *and* $\frac{1}{p}+\frac{1}{q}=1$, *then*

$ab \le \frac{a^p}{p}+\frac{b^q}{q}$ *where a, b are two non-negative reals.*

Proof :

if a = 0 or b = 0 the result is obvious.

Now consider the case in which $a \neq 0$, $b \neq 0$.

Let $f(t) = 1-\lambda + \lambda t - t^{\lambda}$ for $0 < \lambda < 1$.

Then $\quad f'(t) = \lambda - \lambda t^{\lambda-1}$

$\quad f'(t) = 0 \Rightarrow \lambda(1 - t^{\lambda-1}),$

where $\quad \lambda, \lambda - 1 \neq 0$

$$\Rightarrow t = 1$$

$$f''(t) = -\lambda(\lambda - 1)t^{\lambda-2}.$$

At $t = 1$, $f''(t) = -\lambda(\lambda - 1) = +$ ive as $0 < \lambda < 1$.

Thus, $\quad \frac{df}{dt} = 0, \frac{d^2f}{dt^2} >$ at $t = 1$.

This proves that f is minimum at t = 1.

$\therefore \quad f(1) \le f(t)$

This $\quad \Rightarrow 1 - \lambda + \lambda - 1 \le 1 - \lambda + \lambda t - t^{\lambda}$

or $\quad 1 - \lambda + \lambda t - t^{\lambda} \ge 0.$

or $\quad (1 - \lambda) + \lambda t \ge t^{\lambda}$ (Put $\lambda = 1/p$)

or $\quad \left(1-\frac{1}{p}\right)+\frac{t}{p} \ge \frac{t^1}{p}$

But $\frac{1}{p}+\frac{1}{q}=1$

Hence $\quad \frac{1}{q}+\frac{t}{p} \ge \frac{t^1}{p}$. . Put $t = \frac{a^p}{b^q}$.

Then $\quad \frac{1}{q} + \frac{a^p}{pb^q} \ge \frac{a}{b^{q/p}}$

or $\quad \frac{b^q}{q} + \frac{a^p}{p} \ge ab^{q-q/p} = ab.$

or $$\frac{a^p}{p} + \frac{b^q}{q} \geq ab$$

as $$\frac{1}{p} + \frac{1}{q} = 1 \Rightarrow q - \frac{q}{p} = 1$$

HOLDER'S INEQUALITY

If a_i, b_i are non-negative real numbers,

then $$\sum_{i=1}^{n} a_i b_i \leq \left(\sum_{i=1}^{n} a_i^p\right)^{1/p} \left(\sum_{i=1}^{n} b_i^q\right)^{1/q}$$

where $\frac{1}{p} + \frac{1}{q} = 1$ and $p > 1$.

Proof :

Firstly we shall establish a lemma :

Lemma 2:

If $a \geq 0$, $b \geq 0$, $p > 1$, then

$$\frac{a^p}{p} + \frac{b^q}{q} \geq ab$$

where $\frac{1}{p} + \frac{1}{q} = 1.$

By the lemma, $$\frac{A^p}{p} + \frac{B^q}{q} \geq AB \qquad \ldots (1)$$

Put $A = \dfrac{a_i}{\left(\sum_i a_i^p\right)^{1/p}}$, $B = \dfrac{b_i}{\left(\sum_i b_i^q\right)^{1/q}}$

$$\therefore \quad \frac{1}{p} \frac{a_i^p}{\left(\sum_i a_i^p\right)} + \frac{1}{q} \frac{b_i^q}{\left(\sum_i b_i^q\right)} \geq \frac{a_i b_i}{\left(\sum_i a_i^p\right)^{1/p} \left(\sum_i b_i^q\right)^{1/q}}$$

Sum the result from

$i = 1$ to n,

$$\frac{1}{p\left(\sum_i a_i^p\right)} \left(\sum_{i=1}^{n} a_i^p\right) + \frac{1}{q\left(\sum_i b_i^q\right)} \left(\sum_{i=1}^{n} b_i^q\right)$$

$$\geq \frac{\sum_i a_i b_i}{\left(\sum_i a_i^p\right)^{1/p}\left(\sum_i b_i^q\right)^{1/q}}$$

or
$$\frac{1}{p}+\frac{1}{q} \geq \frac{\sum_i a_i b_i}{\left(\sum_i a_i^p\right)^{1/p}\left(\sum_i b_i^q\right)^{1/q}}$$

But $\frac{1}{p}+\frac{1}{q} = 1.$

Hence
$$\left(\sum_{i=1}^{n} a_i^p\right)^{1/p}\left(\sum_{i=1}^{n} b_i^q\right)^{1/q} \geq \sum_{i=1}^{n} a_i b_i$$

Notes: (1) Holder's inequality for complex numbers :

$$\sum_{i=1}^{n} |a_i b_i| \leq \left(\sum_{i=1}^{n} |a_i|^p\right)^{1/p}\left(\sum_{i=1}^{n} |b_i|^q\right)^{1/q}$$

(2) Holder's inequality for integrals :

CAUCHY SCHWARZ INEQUALITY

If a_i, b_i $(i = 1, 2 \ldots n)$ are real or complex numbers, then

$$\left(\int_a^b fg\,dx \leq \left(\int_a^b f^p\,dx\right)^{1/p}\left(\int_a^b g^q\,dx\right)^{1/q}\right)$$

$$\sum_{i=1}^{n} |a_i b_i| \leq \left[\sum_{i=1}^{n} |a_i|^2\right]^{1/2}\left[\sum_{i=1}^{n} |b_i|^2\right]^{1/2}$$

Proof :

If we replace a_i, b_i, p, q by $|a_i|$, $|b_i|$, 2, 2 respectively, then we get this inequality. An independent proof is given below :

Write $A = (a_1, a_2, a_3, \ldots a_n)$, $b = (b_1, b_2, \ldots, b_n)$.

The magnitudes of these vectors are given by

$$\|a\|^2 = \sum_{i=1}^{n} |a_i|^2 \quad \|b\|^2 \sum_{i=1}^{n} |b_i|^2$$

If a = 0 or b = 0, then the inequality reduces to equality.

So we suppose that $a \neq 0$. $b \neq 0$. We know that

Geometric $\leq$ mean Arithmetic mean.

This $\Rightarrow \sqrt{(xy)} \leq \dfrac{x+y}{2}$

Set $\sqrt{x} = \dfrac{|a_i|}{\|a\|}, \sqrt{y} = \dfrac{|b_i|}{\|b\|}$

Then $$\frac{2|a_i|.|b_i|}{\|a\|.\|b\|} \leq \frac{|a_i|^2}{\|a\|^2} + \frac{|b_i|^2}{\|b\|^2}$$

or $$2\sum_{i=1}^{n} \frac{|a_i|.|b_i|}{\|a\|.\|b\|} \leq \frac{\sum|a_i|^2}{\|a\|^2} + \frac{\sum|b_i|^2}{\|b\|^2} = 1+1=2$$

or $$\sum_i |a_i|.|b_i| \leq \|a\|.\|b\|$$

or $$\sum_{i=1}^{n} |a_i b_i| \leq \left(\sum_{i=1}^{n}|a_i|^2\right)^{1/2} \left(\sum_{i=1}^{n}|b_i|^2\right)^{1/2}$$

Note : If a_i , b_i are real numbers, then

$$\sum_i |a_i\, b_i| \leq \left(\sum_{i=1}^{n} a_i^{\,2}\right)^{1/2} \left(\sum_{i=1}^{n} b_i^{\,2}\right)^{1/2}$$

MINKOWSKI INEQUALITY

If $p > 1$ and a_i , b_i are non-negative real numbers, then

$$\left[\sum_{i=1}^{n}(a_i + b_i)^p\right]^{1/p} \leq \left(\sum_{i=1}^{n} a_i^p\right)^{1/p} + \left(\sum_{i=1}^{n} b_i^p\right)^{1/p}$$

Proof :

By Holder's inequality,

$$\sum_{i=1}^{n} a_i + b_i \leq \left(\sum_{i=1}^{n} a_i^p\right)^{1/p} \left(\sum_{i=1}^{n} b_i^q\right)^{1/q} \qquad \ldots (1)$$

with $\dfrac{1}{p} + \dfrac{1}{q} = 1$ ... (2)

By (2), $q\,(p - 1) = p$ ... (3)

$$\sum_{i=1}^{n}(a_i + b_i)^p = \sum_{i=1}^{n}(a_i + b_i)(a_i + b_i)^{p-1}$$

where $\sum\limits_{i=1}^{n} = \sum$

$$= \sum a_i (a_i + b_i)^{p-1} + \sum b_i (a_i + b_i)^{p-1}$$

Using (1) for each member of R.H.S.,

$$\sum (a_i + b_i)^p \leq \left(\sum a_i^p\right)^{1/p} \left[\sum (a_i, b_i)^{q(p-1)}\right]^{1/q}$$

$$+ \left(\sum b_i^p\right)^{1/p} \left[\sum (a_i + b_i)^{q(p-1)}\right]^{1/q}$$

Using (3), $\sum (a_i + b_i)^p \leq \left(\sum a_i^p\right)^{1/p} \left[\sum (a_i + b_i)^p\right]^{1/q}$

$$+ \left(\sum b_i^p\right)^{1/p} \left[\sum (a_i + b_i)^p\right]^{1/q}$$

Dividing by $\left[\sum (a_i + b_i)^p\right]^{1/q}$ we get

$$\left[\sum (a_i + b_i)^p\right]^{1-(1/q)} \leq \left(\sum a_i^p\right)^{1/p} + \left(\sum b_i^p\right)^{1/p}$$

or, $\left[\sum_{i=1}^{m} (a_i + b_i)^p\right]^{1/p} \leq \left(\sum a_i^p\right)^{1/p} + \left(\sum b_i^p\right)^{1/p}$

Note : Minkowski's inequality for integrals

$$\left[\int_a^b (f+g)^p\right] \leq \left(\int_a^b f^p\right)^{1/p} + \left(\int_a^b g^p\right)^{1/p}$$

where $p > 1$.

MINKOWSKI'S INEQUALITY OF NORMS

Let $a = (a_1, a_2, \ldots, a_n)$, $b = b_1, b_2, \ldots, b_n)$ be two n tuples or real or complex numbers, then

$$\|a+b\| \leq \|a\| + \|b\|$$

$$\left[\sum_{i=1}^{n} |a_i + b_i|^2\right]^{1/2} \leq \left(\sum_{i=1}^{n} |a_i|^2\right)^{1/2} + \left(\sum_{i=1}^{n} |b_i|^2\right)^{1/2}$$

Proof:

We have $\|a + b\|^2 = \sum_i |a_i + b_i|^2$,

or $\|a\|^2 = \sum_i |a_i|^2$

$$\|b\|^2 = \sum_i |b_i|^2$$

where $\sum\limits_{i}$ *means* $\sum\limits_{i=1}^{n}$

$$\|a+b\|^2 - \sum_i |a_i + b_i|^2 = \sum_i |a_i + b_i|.|a_i + b_i|$$

$$\leq \sum_i |a_i + b_i|.(|a_i + b_i|)$$

$$\leq \sum_i |a_i + b_i|.\ |a_i| + \sum_i |a_i + b_i|.\ |b_i|$$

$$\leq \sum_i |(a_i + b_i)a_i| + \sum_i |(a_i + b_i)b_i|$$

Applying Cauchy Schwarz inequality to each sigma or R.H.S., we get

$$\|a+b\|^2 \leq \left(\sum_i |a_i + b_i|^2\right)^{1/2} \left(\sum_i |a_i^2|\right)^{1/2}$$

$$+ \left(\sum_i |a_i + b_i|^2\right)^{1/2} . \left(\sum_i |b_i^2|\right)^{1/2}$$

$$= \|a+b\| . \|a\| + \|a+b\| . \|b\|$$

Dividing by $\|a+b\|$, we get

$$\|a+b\| \leq \|a\| + \|b\|$$

Theorem:

If a_i, b_i (i = 1, 2 ... n) are non- negative real numbers and $p > 1$, then

$$\sum_{i=1}^{n}(a_i + b_i)^p \leq \sum_{i=1}^{n} a_i^p + \sum_{i=1}^{n} b_i^p$$

Proof :

Let $f(x) = 1 + x^p - (1 + x)^p$.

Then taking derivative ; it can be shown that

$f(x) \geq 0$ for $x \geq 0$.

or $1 + x^p - (1 + x)^p \geq 0$.

or $(1 + x)^p \leq 1 + x^p$

put $x = a_i/b_i$, we get

$$\left(1 + \frac{a_i}{b_i}\right)^p \leq 1 + \left(\frac{a_i}{b_i}\right)^p$$

or, $(a_i + b_i)^p \leq (a_i)^p + (b_i)^p$

or $$\sum_{i=1}^{n}(a_i + b_i)^p \leq \sum_{i=1}^{n} a_i^p + \sum_{i=1}^{n} b_i^p$$

DEFINITION : HILBERTSPACE

(or l_2-space) : Let H be the set of infinite real sequences

$< x_n >$ s.t. $\sum_{i=1}^{n} x_i^2 < \infty$.

Let x , y ∈ H s. t $x = < x_n >$, $y = < y_n >$. Define a map

$$d : H \times H \to R \text{ s.t. } d(x, y) = \left[\sum_{i=1}^{\infty}(x_i - y_i)^2\right]^{1/2}$$

Aim. d is a metric in H.

(1) To prove that d is well defined map

$$d(x, y) = [\Sigma \{x_i + (-y_i)\}^2]^{1/2}$$
$$\leq (\Sigma x_i^2)^{1/2} + (\Sigma (-y_i)^2)^{1/2}$$

by Minkowski inequality

$$= \Sigma x_i^2 + \Sigma y_i^2 < \infty, \text{ by given condition,}$$

or, $d(x, y) < \infty$,

meaning thereby d s a finite real number. Hence d is well defined map.

(2) $d(x, y) \geq 0$

For x_i , y_i, are real

$\Rightarrow (x_i - y_i)^2 \geq 0$

$\Rightarrow d(x , y) \geq 0$

(3) $d(x , y) = d(y , x)$

For $(x_1 - y_i)^2 = (y_i - x_i)^2$

$\Rightarrow \{\Sigma (x_1 - y_i)^2\}^{1/2} = \{\Sigma (y_i - x_i)^2\}^{1/2}$

$\Rightarrow d(x , y) = d(y , x)$

(4) $d(x , z) \leq d(x, y) + d(y, z)$.

where $z = < z_n > \in H$.

$$d(x,z) = \left[\sum_{i=1}^{\infty}(x_i - z_i)^2\right]^{1/2}$$

$$= [\Sigma \{(x_i - y_i) + (y_i - z_i)\}^2]^{1/2}$$

$\leq [\Sigma (x_i - y_i)^2]^{1/2} + [\Sigma (y_i - z_i)^2]^{1/2}$,

by Minkowski inequality

$= d(x, y) + d(y, z)$.

The above facts prove that d is a metric on H. The space (H, d) is called Hilbert space or l_2-space.

DEFINITION CANTOR'S TENNARY SETS

1. Let 1 = [0, 1]. Divide the closed interval l into three equal parts and remove the middle open interval (1/3 , 2/3), this leaves two disjoint closed intervals [0, 1/3] and [2/3 , 1] each of length 1/3. This is first stage or our construction. We now divide the remaining two closed intervals [0, 1/3] and [2/3 , 1] , each in three equal parts. Delete the middle open interval (1/9, 2/9) and (7/9, 8/9). This is the second stage of our construction. Similarly at the nth stage, 2^{n-1} open intervals are removed each or length $1/3^n$. The points which remain constitute Contor's tennary set. We denote this set by E. Let x = 0. a_1 a_2 a_3 ... (in the scale of three) be any point of E. Then

$$x = \frac{a_1}{3} + \frac{a_2}{3^2} + \frac{a_3}{3^3} + \ldots$$

Hence a_n can not take values 3 or greater than 3.

Hence the only possibility is that $a_n = 0, 1, 2 \ \forall$ n.

At the first step, all the points with $a_2 = 1$ are removed

Hence $a_n = 0$ or $2 \ \forall$ n.

2. To prove that cardinal number of E is c.,

i.e., is uncountable. Recall that

$\{ 0, 2)^N = \{f : N \to \{9, 2\}\}$

i.e., $\{0, 2\}^N$ is the set of functions from N to 0, 2. In otherwords, $\{0, 2\}^N$ is the set of sequence $< x_n >$ where each $x_n = 0$ or 2.

Evidently, the map $0. x_1 x_2 x_3 \ldots \to < x_1 , x_2 \ldots >$ is one one onto map.

This $\Rightarrow \exists$ one-one onto map between E and $\{0, 2\}^N$.

$$\Rightarrow |E| = \left|\{0,2\}^N\right| = 2^a = C$$

$\Rightarrow$ cardinal number of E is c.

Theorem 1:

Let (x, ρ) be a metric space, Show that ∅ and X are open sets in X.

Proof :

Let (X, ρ) be a metric space.

To prove that ∅ and X are open sets .

To prove that ∅ is open in X, it suffices to show that

any $x \in \Phi \Rightarrow \exists\ \varepsilon > 0$, s.t. $S_{\varepsilon\,(x)} \subset \varnothing$.

Since ∅ does not contain any element and hence this condition is automatically fulfilled.

Corresponding to every point $x \in X$, ∃ an open sphere with its centre at x which is contained in X, showing there by X is an open .

Similar Example : *Is the subset [0, 3] open in X = [0, 3] under the metric d given by d (x, y) = | x – y?*

Solution:

Since Φ and X both are open sets in any metric space (X, d).

∴ [0, 3] is open in x = [0, 3] .

Theorem 2:

In a metric space (X, ρ), ∅ and X are closed sets.

Proof :

Let (X, ρ) be a metric space.

To prove that ∅ and X are closed sets.

∵ $D(\varnothing) = \varnothing \subset \varnothing$

∴ $D(\varnothing) \subset \varnothing$

∵ This ⇒ ∅ is a closed set.

All the limiting points of X belong to X. For X is the universal set.

i.e., any $x \in D(X) \Rightarrow x \in X$

∵ $D(X) \subset X$.

This ⇒ X is a closed set.

Theorem 3:

Let (X, d) be a metric space. Show that $F \subset X$, F is closed ⇔ F' is open.

Proof:

Let (X, d) be a metric space .

Let C be a closed subset of X, so that $D(F) \subset F$.

To prove that F' is open in X.

Let X $\in$ F' be arbitrary. Then X $\notin$ F.

$$D(F) \subset F.\ x \notin F \Rightarrow x \notin D(F)$$

$$\Rightarrow (S_{r(x)} - \{x\}) \cap F = \varnothing \text{ for some } r > 0.$$

$$\Rightarrow S_{r(x)} \cap F = \varnothing \qquad \because x \notin F$$

$$\Rightarrow S_{r(x)} \subset X - F$$

$$\Rightarrow S_{r(x)} \subset F'.$$

$\therefore$ Given x $\in$ F', $\exists$ any an open sphere $S_{r(x)}$ s.t.

$$S_{r(x)} \subset F'.$$

By definition, this proves that F' is open.

Conversely suppose that F' is open in X.

To prove that F is closed in X.

Let x $\in$ F' be arbitrary, then x $\notin$ F.

$\because$ F' is open, $\exists$ r $\subset R^+$ s.t., $S_{r(x)} \subset F'$

$$\text{This} \Rightarrow S_{r(x)} \cap F = \varnothing$$

$$\Rightarrow (S_{r(x)} - \{x\}) \cap F = \varnothing$$

$$\Rightarrow x \notin D(F).$$

Thus, any x $\in$ F' $\Rightarrow$ x $\notin$ D (F).

i.e., any x $\in$ X – F

$$\Rightarrow x \in X - D(F)$$

This $\Rightarrow$ X – F $\subset$ X – D (F)

or D (F) $\subset$ F.

This $\Rightarrow$ F is closed.

Theorem 4:

In any metric space (X, d) each open sphere is an open set.

Proof:

Let (X, d) be a metric space. Let $S_{r0\,(x0)}$ be an open sphere in X.

To prove that $S_{r0\,(x0)}$ is an open set.

Let x $\in S_{r0\,(x0)}$ be arbitrary, then $d(x, x_0) < r_0$

Write $r = r_0 - d(x, x_0)$, ...(1)

By definition, $S_{r0\,(x0)} = \{y \in X : d(y, x_0) < r_0\}$

$S_{r(x)} = \{y \in X : d(y, x) < r\}$.

We claim $S_{r\,(x)} \subset S_{r0\,(x0)}$

Let $y \in S_{r\,(x)}$ be arbitrary.

Then $d(x, y) < r$

$d(y, x_0) \leq d(y, x) + d(x, x_0)$

$< r + d(x, x_0) = r_0.$

$\therefore$ $d(y, x_0) < r_0$ on using (1)

This $\Rightarrow$ $y \in S_{r0\,(x0)}$

and $y \in S_{r\,(x)}$

$\Rightarrow$ $y \in S_{r0\,(x0)}$

This $\Rightarrow$ $S_{r\,(x)} \subset S_{r0\,(x0)}$

Thus we have shown that

given any $x \in S_{r0\,(x0)}$, $\exists\, r > 0$ s.t. $S_{r\,(x)} \subset S_{r0\,(x0)}$.

By definition, this proves that $S_{r0\,(x0)}$ is an open set.

Theorem 5:

In any metric space, any closed sphere is a closed set.

Proof:

Let $Sr_{0\,[x0]}$ denote a closed sphere in a metric space (X, d).

To prove that $Sr_{0\,[x0]}$ is a closed set.

For this we must show that $S'\, r_0\, [x_0]$ is open in X.

Let $x \in S'\, r_{0\,[x0]}$ be arbitrary,

$x \in S'\, r_{0\,[x0]} \Rightarrow x \notin S'\, r_{0\,[x0]}$

$\Rightarrow d(x, x_0)\, , <\, , \neq r_0,$

$\because Sr_{0\,[x0]} = \{ y \in X : d(y, x_0) \leq r_0 \}$

$\Rightarrow d(x, x_0) > r_0$

$\Rightarrow d(x, x_0) - r_0 > 0$

$\Rightarrow r > 0$, on taking $r = d(x, x_0) - r_0$ ...(1)

We claim $S_{r\,(x)} \subset S'\, r_{0\,[x0]}.$

Let $y \in S_{r\,(x)}$ be arbitrary, so that, $d(y, x) < r$.

$\therefore$ $d(x, x_0) \leq d(x, y) + d(y, x_0).$

$\therefore$ $d(y, x_0) \geq d(x, x_0) - d(x, y).$

$> d(x, x_0) - r = r_0$ on using (1)

$\therefore \quad d(y, x_0) > r_0$

$\Rightarrow y \notin Sr_{0\,[x0]}$

Thus, any $\quad y \in S_{r\,(x)} \Rightarrow y \in S'r_{0\,[x0]}$

This $\quad \Rightarrow S_{r\,(x)} \subset S'r_{0\,[x0]}$

$\therefore$ Given any $x \in S'r_{0\,[x0]}$, $\exists\, r > 0$ s.t. $S_{r\,(x)} \subset S'r_{0\,[x0]}$

This proves that $S'r_{0\,[x0]}$ is open in X.

Theorem 6:

Let (X, d) be a metric space and let $G \subset X$ be an arbitrary set. Show that G is open $\Leftrightarrow$ it is a union of open spheres.

Proof:

Let (X, d) be a metric space. Let $G \subset X$ be an open set.

To prove that G is a union of open spheres.

If $G = \varnothing$, the proof is trivial,

If $G \neq \varnothing$, then by definition of an open set

any $\quad x \in G \Rightarrow \exists\, r \in R^+$ s.t. $S_{r\,(x)} \subset G$ ...(1)

In view of this statement we are justified to write.

$$G = \bigcup_i \{S_{ri(xi)} : x_i \in G\}$$

where each $r_i \in R^+$ is s.t. $S_{ri(xi)}$ satisfies the condition (1).

Conversely suppose that (X, d) is a metric space. Let $G \subset X$ be expressible as a union of open spheres, *i.e.*, $G = \bigcup_i S_{ri(xi)}$.

To prove that G is an open set.

Let $x \in G$ be arbitrary.

Then x is a centre of some open sphere, say.

$$S_{(x,\,r)} \in \{S_{ri(xi)}\}$$

Then $\quad S_{r\,(x)} \subset \bigcup_i S_{ri\,(xi)} = G$,

or $\quad S_{r\,(x)} \subset G$,

$\therefore$ any $\quad x \in G \Rightarrow \exists\, r \in R^+$ s.t. $S_{r\,(x)} \subset G$.

By definition this proves that G is an open set.

Theorem 7:

Let (X, d) be a metric space. Prove that

(i) an arbitrary union of open sets is open in X.

(ii) any finite intersection of open sets is open.

Proof:

Let (X, d) be a metric space.

Let $G_i \subset X$ be open X $\forall$ i $\in$ N

Let $G = \bigcap_{i=1}^{\infty} G_i$, $G = \bigcap_{i=1}^{n} G_i$

To prove that

(i) G is open in X.

(ii) G is open in X.

(i) By Theorem 7, each G_i is a union of open spheres and hence G is the union of all open spheres arising in this way.

Hence g is open in X.

(ii) If G = Φ, then G is open in X. For Φ is open in X.

If $G \neq \Phi$, then $\exists\, x \in \bigcap_{i=1}^{n} G_i$ so that

$x \in G_i$ for $1 \leq i \leq n$.

$x \in G_i$, G_i is open

$\Rightarrow \quad \exists\, r_i \in R^+$ s.t. $S_{r\,(x)} \subset G_i$ ...(1)

This is true for each i (= 1, 2, ..., n).

Take $r = \min \{r_1, r_2 ..., r_n\}$

Then $r \leq r_i$ for $1 \leq i \leq n$.

This $\Rightarrow S_{r\,(x)} \subset S_{ri\,(x)}$ for $1 \leq i \leq n$

$\Rightarrow S_{r\,(x)} \subset S_{ri\,(x)} \subset G_i$ by (1)

$\Rightarrow S_{r\,(x)} \subset G_i$

$\Rightarrow S_{r\,(x)} \subset G = \bigcap_{i=1}^{n} G_i = G \Rightarrow S_{r\,(x)} \subset G$

$\therefore$ any $x \in G \Rightarrow \exists\, r \in R^+$ s.t. $S_{r\,(x)} \subset G$

By definition, this $\Rightarrow$ G is open in X.

Remark: The intersection of an infinite number of open sets may or may not be open. If we take an open set, $A_n = (-1/n\,,\, 1/n)$, then

$$\bigcap_{n=1}^{\infty} A_n = \{0\} = \text{closed set}]$$

Theorem 8:

In any metric space (X, d) , show that

(i) an arbitrary intersection of closed sets is closed.

(ii) any; finite union on closed sets is closed.

Proof:

Let (X, d) be a metric space

Let $F_i \subset X$ be a closed set $\forall$ i $\in$ N

Let $F = \bigcap_{i=1}^{\infty} F_i$, $F = \bigcup_{i=1}^{n} F_i$

To Prove that F and F are closed sets in X.

By theorem 4, F_i is closed $\Leftrightarrow$ F_i' is open.

By theorem 8, $\bigcup_{i=1}^{\infty} F_i', \bigcap_{i=1}^{n} F_i'$ are open in X. ...(1)

By, De Morgan's law, $\bigcup_{i=1}^{\infty} F_i', = (\bigcap_{i=1}^{\infty} F_i) = F'$

and $\bigcap_{i=1}^{\infty} F_i' = (\bigcup_{i=1}^{\infty} F_i) = F'$

From (1), F' , F' are open in X

i.e., F, F are closed in X. [by Theorem 4]

Remark: The union of an infinite number of closed sets may or may not be closed.

Let $F_n = \left[0, \frac{n}{n+1}\right]$

Then $\bigcup_{i=1}^{\infty} F_n = [0, 1)$ = semi open set

$\therefore$ $\bigcup_{n=1}^{\infty} F_n$ is not closed, though each F_n is closed.

Examples:

To find the boundary of set of integers Z and set of rational Q.

$Z^o = \cup \{ G \subset R : G \text{ is open and } G \subset Z \} = \phi$

(1) For every subset of R contains fractions also.

Similarly $(R - Z)^o = \phi$

$b(Z) = R - Z^o \cup (R - Z)^o = R - \phi \cup \phi = R$

or, $b(Z) = R$

Similarly $b(Q) = R$

(2) Give two examples of limit points.

(i) If $A = \left\{1+\frac{1}{n} : n \in N\right\}$,

$$A = \left\{2, \frac{3}{2}, \frac{4}{3}, \frac{5}{4}, \frac{6}{5}, \ldots\right\}, \text{ then}$$

1 is limit point of A for $\lim_{n\to\infty}\left(1+\frac{1}{n}\right) = 1.$

(ii) If $A = \left\{1, \frac{1}{2}, \frac{1}{3}, \frac{1}{4}, \ldots\right\} = \left\{\frac{1}{n} : n \in N\right\}$,

then 0 is the limit point of A.

For $\lim_{n\to\infty} \frac{1}{n} = 0.$

(3) Since N has no limit points and so D (N) = ϕ*. For every open sphere centred at any point of N other than that point.*

$$\overline{N} = N \cup D(N) = N \cup \phi = N$$

or $\overline{N} = N$ $\therefore$ *N is closed.*

so that N' is open and so $(N')^o = N'$.

$$ext(N) = (N')^o = N'$$

$$b(N) = [N^o \cup ext(N)]' = [\phi \cup ext(N)']$$

$$= (N')' = N$$

Note that N is non-dense set.

For $(\overline{N})^o = N^o = \phi$

Example:

Define non where dense set and give an example.

(4) Consider the set R. Evidently $Q \subset R$*. Every real number is a limit point so that rational limit points belong to the set Q.*

It means that $D(R) = R$, $D(Q) = R$

$$\because \quad \overline{A} = A \cup D(A)$$

$$\therefore \quad \overline{Q} = Q \cup D(Q) = Q \cup R = R$$

This proves that the closure of the set of rational numbers w.r.t. usual metric no R is R.

Thus $Q \subset R$ *, Q is countable and* $\overline{Q} = R$.

This proves that r with usual metric is separable.

$Q^o = \cup \{G \subset R : G \text{ is open}, G \subset Q = \phi$.

Since every subset of R contains rationals and irrationals. Hence their exists no open set.

$$G \subset R \text{ s.t. } G \subset Q.$$

Thus, $Q^o = \phi.$

Similarly $ext\ (Q) = (R - Q)^o = \phi.$

$int\ (\overline{Q}\ ) = int\ (R) = R \neq \phi.$

$\therefore$ *Q is some where dense set in R.*

$$b\ (Q) = R - Q^o \cup (R - Q)^o = R - \phi \cup \phi = R$$

(5) If $A = (a,\ b)$ *or* $[a,b)$ *or* $(a,\ b]$ *or* $[a,\ b]$,

then in every case,

$\overline{A} = [a,\ b],\ A^o = (a,\ b),\ b(A) = \{a,\ b\}.$

(6) If $A = [0,\ 1)$, *then* $(\ \overline{A}\ )^o = [0,\ 1]^o = (0,\ 1) \neq \phi$.

and so A is somewhere dense set in R.

(7) To show that N and Z are non-dense sets in R relative to the usual metric.

$$D\ (N) = \phi.$$

For if a is any real number, then any nbd $= (\ a - \varepsilon\ ,\ a + \varepsilon\ ) \subset R$ *does not contain any point of N other than a*

$$Z = \{n : n \varepsilon N\} \cup \{0\} \cup \{-n : n \varepsilon N\}$$

$$D\ (Z) = \phi \cup \phi \cup \phi = \phi$$

or, $D\ (Z) = \phi \subset Z$

or, $D\ (Z) \subset Z.$

$\therefore$ *Z is a closed set. Similarly N is a closed set.*

For $D\ (A) \subset A \Leftrightarrow A$ *is closed.*

$\therefore$ $\overline{Z} = Z$ *and* $\overline{N} = N$

$Int\ \overline{Z} = Int\ (Z) = \cup \{G \subset R : G \text{ is open and } G \subset Z\} = \phi$

For any open subset $(a_1,\ a_2) \subset R$ *contains fractions also.*

Now $Int\ (\overline{Z}\ ) = (\overline{Z}\ )^o = \phi$.

Similarly $(\overline{N})^o = \phi$.

$\therefore$ *Z and N arc nowhere dense subsets of R.*

DEFINITION OF NEIGHBOURHOOD

Let (X, d) be a metric space and $x \in X,\ A \subset X$.

A subset A of X is called a neighbourhood (nbd) of x if $\exists$ open sphere $S_{r(x)}$ s.t. $S_{r(x)} \subset A$.

This means that A is nbd of a point x iff x is an interior point of A.

From the definition of nbd it is clear that :

(1) Every superset of a nbd of a point is also a nbd.

(2) Every open sphere $S_{r(x)}$ is a nbd of x.

(3) Every closed sphere $S_{r[x]}$ is a nbd of x.

(4) Intersection of two nbds of the same point is again a nbd of that point.

(5) A set is open if it contains a nbd of each of its points.

(6) Nbd of a point need not be an open set.

Theorem 9:

A subset in a metric space is open iff it is a nbd of each of its point.

Proof:

Let A be a subset of a metric space (X, d).

Step I : Given A is and of each of its points.

Aim : A is an open set. Recall that a set N is called nbd of a point $x \in X$ if $\exists$ open set $G \subset X$ s.t. $x \in G \subset N$.

X s.t. $p \in G_p \subset A$.

It is true $\forall$ $p \in A$.

Take $A = \cup \{G_p : p \in G_p, G_p$ is open set, $G_p \subset A\}$

= An arbitrary union of open sets

= open set.

$\therefore$ A is an open set.

Step II : Let A be an open subset of X.

Aim : A is a nbd of each of its points. By assumption, we can write

$$p \in A \subset A \ \forall \ p \in A$$

This $\Rightarrow$ A is a nbd of each of its points.

Theorem 10:

A subset A of metric subspace (Y, d) of a metric space (X, d) is open in Y iff $\exists$ an open set G in (X, d) s.t. $A = Y \cap G$.

Proof:

Let $A \cap Y$ and let (Y, d) be a subspace of a metric space (X, d).

Then $\qquad A \subset Y \subset X \qquad$ ...(1)

Step I : Let A be open in Y. ...(2)

Aim : $A = Y \cap G$ for some open set G in X. By def. of open set,

$\forall\ y \in A \Rightarrow\ \exists$ open sphere $S_{1\,(Y,\,r)}$ in s.t.

$$S_{1\,(Y,\,r)} \subset A \qquad ...(3)$$

Then A is expressible as

$$A = \cup \{S_{1\,(Y,\,r)} : y \in A\} \qquad ...(4)$$

Suppose S(Y, r) is an open sphere in X.

Then $\quad S_{1\,(Y,\,r)} = Y \cap S(Y, r)$

Now (4) is expressible as

$$A = \cup \{S_{(Y,r)} \cap Y : y \in A\}$$

$$= Y \cup [S_{(Y,r)} : y \in A\}$$

or, $\quad A = Y \cap G,$

where $\quad G = \cup \{S_{(Y,r)} : y \in A\}$

= arbitrary union of open spheres in X

= open set in X.

$\therefore\ A = Y \cap G$, where G is open in X.

For each open sphere is an open set.

Step II : Let $A = Y \cap G$ ...(5)

where G is an open set in X.

Aim : A is open in Y.

By def. of open set in X,

$\forall\ x \in G \subset\ \Rightarrow \exists$ open sphere $S_{(x,\in)}$ in X. s.t.

$$S_{(x,\in)} \subset G \qquad ...(6)$$

Write $\quad S_{1(x,\hat{I})} = Y \cap S_{(x,\hat{I})}$ ...(7)

Then $S_{1(x,\in)}$ is an open sphere in Y.

By (6), $\quad S_{(x,\in)} \cap Y \subset G \cap Y$

Using (7), we get $S_{1(x,\in)} \subset G \cap Y = A$, by (5)

or, $\quad x \in S_{1(x,\in)} \subset A$

This $\Rightarrow$ A is open in Y, be def. of open set.

Corollary : *The necessary and sufficient condition that a subset A of a metric subspace (Y, d) of a metric space (X, d) be closed in (Y, d) is that $\exists$ a set F closed in (X, d) s.t.*

$$A = Y \cap F.$$

Proof:

Firstly we shall prove a lemma ·

A is open in Y $\Leftrightarrow$ A Y $\cup$ G, G is open in X.

Now we come to the proof of the actual theorem.

F is closed in X $\Rightarrow$ X – F is open in X.

A is closed in Y $\Leftrightarrow$ Y – A is open in Y.

Be the lemma,

$\therefore \quad Y - A = Y \cap (X - F)$

or, $Y - Y \cap (X - F) = A$

or, $Y - (Y \cap X - Y \cap F) = A$

or, $Y - (Y - Y \cap F) = A$

or, $Y \cap F = A$

Theorem 11:

If A_1 is an open set in (X_1, d) and A_2 is an open set in (X_2, d_2), then $A_1 \times A_2$ is open in the product metric space (X,d), where $X = X \times X_2$.

Proof:

Let $x = (x_1, x_2) \in A_1 \times A_2$ be arbitrary, then $x_1 \in A_1, x_2 \in A_2$. By def. of open set,

$\exists$ open spheres $S(x_1, r_1)$, $S(x_2, r_2)$ s.t.

$$\left.\begin{array}{l} S(x_1, r_1) \subset A_1 \\ S(x_2, r_2) \subset A_2 \end{array}\right\} \quad \ldots(1)$$

Let $r = \min. \{r_1, r_2\}$, then $r \le r_1, r \le r_2$.

$\therefore \quad S(x_1, r) \subset S(x_1, r_1) \subset A_1$

$\quad S(x_2, r) \subset S(x_2, r_2) \subset A_2$

This $\quad \Rightarrow S(x_1, r) \times S(x_2, r_1) \subset A_1 \times A_2$

$\quad \Rightarrow x = (x_1, x_2) \in S(x, r)$

$\quad \Rightarrow S(x_1, r) \times S(x_2, r_1) \subset A_1 \times A_2$

$\quad \Rightarrow x \in S(x, r) \subset A_1 \times A_2.$

It is true $\quad \forall\ x \in A_1 \times A_2$

$\therefore \quad A_1 \times A_2$ is open subset of (X, d).

Here S (x, r) is an open sphere in (X, d).

Theorem 12:

Let A be a subset of a metric space (X, d). Then $x_0 \in X$ is a limit point of A iff every open sphere centred at x_0 contains an infinitely many; points of A, other than x_0.

Proof:

Let a be a subset of a metric space (X, d) and let $x_0 \in X$.

Step I : Given x_0 is a limit point of A.

Aim. $S_{r\,(x0)}$ contains an infinite number of points of A other than x_0 ...(*)

By def. of limit point,

$$(S_{r\,(x0)}) - \{x_0\}) \cap A \neq \phi \qquad ...(1)$$

for every open sphere centred at x_0.

Suppose (*) is not true.

Then $S_{r\,(x0)}$ contains a finite number of points

$$x_1, x_2, \ldots, x_n, \in A \qquad ..(2)$$

Let $r = \min \{d(x_0, x_i) : = 1, 2. \ldots, n\}$

and $S_{r(x0)} = \{x \in X : (x, x_0) < r_0\}$

Evidently $(S_{r\,(x0)}) - \{x_0\}) \cap A = \phi$

This $\Rightarrow x_0$ is not a limit point of A.

This is contrary to what is given . This contradiction arises by assuming that $S_{r(x0)}$ contains a finite number of points of A other than x_0. Hence the required result follows.

Step II: Given $S_{r(x0)}$ contains an infinite number of points of A other than x_0.

Aim. x_0 is a limit point of A.

By assumption, $(S_{r\,(x0)}) - \{x_0\}) \cap A \neq \phi$

This $\Rightarrow x_0$ is a limit point of A.

Corollary : *If p is a limit point of A in a metric space, then every nbd N(p) contains infinitely many points of A.*

Or

Every open sphere centred at a limit point of a sat in a metric space contains infinitely many points of the set.

Hint : Open sphere $S_{r(p)}$ can be taken in place of $N_{(p)}$. Here write step I of the above Theorem 13.

Theorem 13:

Let (X, d) be metric space and $A \subset X$. Show that

$$\overline{A} = \{x \in X : d(x, A) = 0\}.$$

Proof:

Let (X, d) be metric space and $A \subset X$.

We claim $\overline{A} = \{x \in X : d(x, A) = 0\}$

Suppose $d(x, A) = \in > 0$

Then (i) $x \notin A$ and

(ii) $(S_{\in/2}(x) - \{x\}) \cap A = \Phi$

Evidently (ii) $\Rightarrow x \notin D(A)$.

Thus, $d(x, A) = \notin \Rightarrow x \notin A, x \notin D(A)$

$\Rightarrow x \notin A \cup D(A) = \overline{A}$

$\Rightarrow x \in \overline{A}$

Finally, $d(x, A) = \varepsilon \Rightarrow x \notin \overline{A}$

Again $d(x, A) = 0 \Rightarrow$

(iii) $x \in A$,

(iv) $(S_{r(x)} - \{x\}) \cap A \neq \Phi$

(iv) $\Rightarrow x \in D(A)$

Finally, $d(x, A) = 0 \Rightarrow x \in A$ or $x \in D(A)$

$\Rightarrow x \in A \cup D(A) = \overline{A} \Rightarrow x \in \overline{A}$

Thus, we have shown that

$d(x, A) = 0 \Rightarrow x \in \overline{A}$

and $d(x, A) = \varepsilon > 0 \Rightarrow x \notin \overline{A}$

Furthermore, $d(x, A) = \geq 0$.

Consequently the required result follows :

Theorem 14:

Let (X, d) be a metric space. A subset A of X is closed if given any $x \in X - A$, $d(x, A) \neq 0$.

Proof :

Let (X, d) be a metric space and $A \subset X$ be an arbitrary closed set.

To prove that

given any $x \in X - A$, $d(x, A) \neq 0$

A is closed $\Rightarrow X - A$ is open.

By definition of open set,

any $x \in X - A \Rightarrow \exists\, r \in R^+$ s.t. $S_{r(x)} \subset X - A \Rightarrow S_{r(x)} \cap A = \Phi$

$\Rightarrow d(x, A) \geq r \Rightarrow d(x, A) \neq 0$.

Conversely let A be any subset of a metric space (X, d). Let

any $x \in X - A \Rightarrow d(x, A) \neq 0$

To prove that A is closed.

Let $x \in X - A$ be arbitrary so that, by assumption,

$$d(x, A) = r \neq 0 \Rightarrow S_{r(x)} \cap A = \phi$$

$\Rightarrow \quad S_{r(x)} \subset X - A$

$\therefore \quad x \in X - A \Rightarrow \exists\, r \in R^+$ s.t. $S_{r(x)} \in X - A$.

By definition, this $\Rightarrow X - A$ is open

$\Rightarrow$ A is closed.

Theorem 15:

Let (X, ρ) be a metric space.

Define $\quad d(x, y) = \dfrac{\rho(x, y)}{1 + \rho(x, y)} \ \forall\, x, y, \in X.$

Then (X, y) is metric space.

Proof:

Let (X, ρ) be a metric space

Let $d(x, y) = \dfrac{\rho(x, y)}{1 + \rho(x, y)} \ \forall\, x, y, \in X.$

To prove that (X, d) is a metric space.

Let x, y, z, $\in$ X be arbitrary.

$\because \quad (X, \rho)$ is a metric space.

$\therefore$ (i) $\quad \rho(x, y) \geq 0$

(ii) $\quad \rho(x, y) = \rho(y, x)$

(iii) $\quad \rho(x, y) = 0$ iff $x = y$

(iv) $\quad \rho(x, y) + \rho(y, z) \geq \rho(x, z)$.

Now (i)' $d(x, y) \geq 0$ in accordance with (i).

(ii)' $d(x, y) = d(y, x)$.

For $d(y, x) = \dfrac{\rho(y, x)}{1 + \rho(y, x)}$

$$= \frac{\rho(x, y)}{1 + \rho(x, y)} = d(x, y) \text{ in accordance with (ii)}$$

(iii)' d (x, y) = 0, iff x = y

$$d(x, y) = 0 \Leftrightarrow \frac{\rho(x, y)}{1 + \rho(x, y)} = 0$$

$$\Leftrightarrow \rho(x, y) = 0 \Leftrightarrow x = y, \text{ according to (iii)},$$

(iv)' d (x, y) + d (y, z) ≥ d (x, z).

For $d(x, y) + d(y, z) = \dfrac{\rho(x, y)}{1 + \rho(x, y)} = \dfrac{\rho(y, z)}{1 + \rho(y, z)}$

$$\geq \frac{\rho(x, y)}{1 + \rho(x, y) + \rho(y, z)} + \frac{\rho(y, z)}{1 + \rho(y, z) + \rho(x, y)}$$

$$\geq \frac{\rho(x, y) + \rho(y, z)}{1 + \rho(x, y) + \rho(y, z)}$$

$$\geq \frac{\rho(x, z)}{1 + \rho(x, z)}$$

$$= d(x, z)$$

i.e., d (x, y) + d (y, z) ≥ d (x, z).

The results (i)', (ii)', (iii)', (iv)' taken together prove the required result.

Theorem 16:

Let (X, y) be a metric space and let m a positive real number. Show that there exists a metric d' for X such that the metric space (X, y) is bounded with d(X) ≤ m.

Proof:

Let (X, y) be a metric space and m > 0.

we define $d(x, y) = \dfrac{md(x,y)}{1 + d(x,y)} \; \forall \; x, y \in X.$

To prove that (X) (X, d) is a metric space and m > 0.

Let x, y, x ∈ X be arbitrary.

d is metric ⇒ d (x, y) ∈ R

Also m ∈ R

$$\Rightarrow d'(x, y) \frac{md(x,y)}{1+d(x,y)} \in R.$$

(X, d) is metric space ⇒

(i) d (x, y) ≥ 0

(ii) d (x, y) = d (y, x)

(iii) d (x, y) = 0 iff x = y

(iv) d (x, y) + d (y, z) ≥ d (x, y).

(1) Evidently d' (x, y) ≥ 0, according to (i) as m > 0.

(2) d' (x, y) = d' (y, x)

For $d'(y, x) \frac{md(y,x)}{1+d(y,x)}$

$$= \frac{md(xy)}{1+d(x,y)} \text{, according to (ii)}$$

$$= d'(x, y)$$

(3) d' (x, y) = 0 iff x = y.

For $d'(x, y) = 0 \Leftrightarrow \frac{md(xy)}{1+d(x,y)} = 0$

$$\Leftrightarrow md(x, y) = 0$$

$$\Leftrightarrow d(x, y) = 0$$

$$\Leftrightarrow x = y \text{, according to (iii).}$$

(4) d' (x, y) + d' (y, z) ≥ d' (x, z).

For $d'(x, y) + d'(y, z) = \frac{md(x,y)}{1+d(x,y)} + \frac{md(y,z)}{1+d(y,z)}$

$$= m\left[\frac{d(x,y)}{1+d(x,y)} + \frac{d(y,z)}{1+d(y,z)}\right]$$

$$\geq m\left[\frac{d(x,y)}{1+d(x,y)+d(y,z)} + \frac{d(y,z)}{1+d(x,y)+d(y,z)}\right]$$

$$= m\left[\frac{d(x,y)+d(y,z)}{1+d(x,y)+d(y,z)}\right] \geq \frac{md(x,z)}{1+d(x,z)}$$

$$= d'(x, z)$$

The results (1), (2), (3) and (4) taken together prove that (X, d') is metric space. Remains to prove that d' is bounded.

$$d'(x, y) = \frac{md(x,y)}{1+d(x,y)} = \frac{m}{1+\frac{1}{d(x,y)}}$$

$d'(x, y) \leq m \ \forall \ x, y, \in X.$

This $\Rightarrow$ (x, d') is a bounded metric space.

Remark: If $d'(x, y) = \dfrac{d(x,y)}{1+d(x,y)}$

Then $d'(x, y) \leq 1$ so that (X, d') is bounded.

Theorem 17:

If A and B are subsets of a metric space (X, d), then

(i) $(A \cap B)^o = A^o \cap B^o$.

(ii) $A^o \cup B^o \subset (A \subset B)^o$.

Also show that $A^o \cup B^o \neq (A \cup B)^o$ *by means of an example.*

Proof:

(i) Any $x \in (A \cap B)^o$

$\Rightarrow$ x is an interior point of $A \cap B$

$\Rightarrow S_r(x) \subset A \cap B$

$\Rightarrow S_r(x) \subset A \cap B \subset A, S_r(x) \subset A \cap B \subset B$

$\Rightarrow S_r(x) \subset A, S_r(x) \subset B$

$\Rightarrow x \in A^o, x \in B^o$

$\Rightarrow x \in A^o \cap B^o$

$\therefore \quad (A \cap B^o) \subset A^o \cap B^o.$

Conversely, any $x \Rightarrow A^o \cap B^o$. ...(1)

$\Rightarrow x \in A^o$ and $x \in B^o$

$\Rightarrow \exists$ open spheres $S_{r1}(x), S_{r2}(x)$, s.t.

$S_{r1}(x) \subset A, S_{r2}(x) \subset B$

Choose $r = \min.(r_1, r_2)$

$\Rightarrow S_r(x) \subset S_{r1}(x) \subset A, S_r(x) \subset S_{r2}(x) \subset B$

$\Rightarrow S_r(x) \subset A, S_r(x) \subset B$

$\Rightarrow S_r(x) \subset A \cap B$

$\Rightarrow x \in (A \cap B)^o$

$\therefore \quad A^o \cap B^o \subset (A \cap B)^o$ ... (2)

Combining (1) and (2), we get the result (i)

(ii) Any $x \in A^o \cup B^o$

$$\Rightarrow x \in A^o \text{ or } x \in B^o.$$

$$\Rightarrow S_{r1}(x) \subset A \text{ or } S_{r2}(x) \subset B$$

Choose the two spheres s.t. one is contained in the other so that $r_2 \geq r_1$ and $S_{r1}(x) \cup S_{r2}(x) = S_{r2}(x)$.

$$\Rightarrow S_{r2}(x) = S_{r1}(x) \cup S_{r2}(x) \subset A \cup B$$

$$\Rightarrow S_{r2}(x) \subset A \cup B$$

$$\Rightarrow x \in (A \cup B)^o$$

$\therefore \quad A^o \cup B^o \subset (A \cup B)^o$.

Example :

Let $A = [0, 1]$, $B = [1, 2]$

Then $\quad A^o = (0, 1)$, $B^o = (1, 2)$.

$$A^o \cup B^o = (0, 1), \cup (1, 2)$$

$$= \{x \in R : 0 < x < 1 \text{ or } 1 < x < 2\}$$

$\because \quad (A \cup B) = [0, 2]$

$\therefore \quad (A \cup B)^o = (0, 2)$.

Evidently $\quad (A \cup B)^o \neq A^o \cup B^o$.

Theorem 18:

Let (X, d) be a metric space and let

$$d'(x, y) = \min\{1, d(x, y)\}.$$

Then d' is a metric for X.

Proof:

Let (X, d) be a metric space so that

(i) $d(x, y) \geq 0$

(ii) $d(x, y) = d(y, x)$

(iii) $d(x, y) = 0$ iff $x = y$

(iv) $d(x, y) + d(y, x) \geq d(x, z)$, where $x, y, z \in X$.

Let $d'(x, y) = \min, \{1, d(x, y)\}$.

Then $\quad d'(x,y) = \begin{cases} 1 \\ \text{or } d(x,y) \end{cases}$ ...(1)

(2) $d'(x, y) \geq 0$, according to (1) and (i),

(3) d' (x, y) = d' (y, x).

For if d' (x, y) = 1, then d' (y, x) = 1 so that d' (x, y) = d' (y, x).

Again if d' (x, y) = d (x, y), then d' (y, x) = d (y, x) using (ii) and combining the last two, we get

$$d'(x, y) = d'(y, x)$$

$\therefore$ In either case, d' (x, y) = d' (y, x)

(4) d' (x, y) = 0 iff x = y.

For d' (x, y) = 0 $\Leftrightarrow$ d (x, y) = 0 according to (1)

$\Leftrightarrow$ x = y according to (iii).

(5) To prove d' (x, y) + d' (y, z) $\geq$ d' (x, z)

If d' (x, y) = 1, then d' (x, z) = 1, d' (y, z) = 1 and

so d' (x, y) + d' (y, z) > d' (x, z)

If we take d' (x, y) = d (x, y), then

d' (y, z) = d (y, z),

d' (x, z) = d (x, z)

and so

$$d'(x, y) + d'(y, z) = d(x, y) + d(y, z)$$
$$\geq d(x, z) = d', (x, z) \text{ d by (iv)}$$

or, d' (x, y) + d' (y, z) $\geq$ d' (x, z)

$\therefore$ (5) is true in both cases.

From what has been it follows that d' is a metric on X.

Similar Problem : Show that if d is a pseudo metric on X and e(x,y) = min {1, d (x, y)}, then e is also a pseudo metric.

Theorem 19:

In any metric space, prove that difference between the two sides of a triangle is less than the third side.

Or

Let (X, d) be a metric space and let x, y, z be any three points of X. Then $|d(x, z) - d(y, z)| \leq d(x, y)$

Proof :

Let x, y, z denote the co-ordinates of vertices of a triangle.

If we show that $|d(x, z) - d(y, z)| \leq d(x, y)$, the result will follow.

We know that

$$d(x, y) + d(y, z) \geq d(x, z)$$

Similarly $\quad d \quad (y, x) + d(x, z) \geq d(y, z)$

From Which $d(x, y) \geq d(x, z) - d(y, z)$ ...(1)

and $\quad d(y, x) \geq -[d(x, z) - d(y, z)]$

The last gives $-d(x, y) \leq d(x, z) - d(y, z)$ ...(2)

For $d(x, y) = d(y, x)$.

Combining (1) and (2),

$$-d(x, y) \leq d(x, z) - d(y, z) \leq d(x, y)$$

This $\quad \Rightarrow |d(x, z) - d(y, z)| \geq d(x, y)$.

Theorem 20:

Let (X, d) be a metric space, then $\forall x, y, x', y'$, *in X,* $|d(x, y) - d'(x' y')| \leq d(x, x') + d(y, y')$.

Proof:

For proving the required result, we make repeated use of "triangle inequality."

$$d(x, y) \leq d(x, x') + d(x', y)$$

$$\leq d(x, + d(x', y') + d(y',$$

or, $\quad d(x, y) - d(x', y') \leq d(x, + d(y',$ ...(1)

as $\quad d(y, = d(y',$

Again $\quad d(x', y') \leq d(x', + d(x, y')$

$$\leq d\ [illegible]', + d(x, y) + d(y,$$

Using the fact that

$d(a, b) = d(b, a)$,

we get $\quad d(x', y') - d(x, y) \leq d(x, + d(y,$

or, $\quad d(x, y) - d(x', y') \geq -[d(x, + d(y,$ ...(2)

Combining (1) and (2), we get,

$-[d(x, + d(y, \leq d(x, y) - d(x', y') \leq d(x, + d(y,$

or, $\quad |d(x, y) - d(x', y')| \leq d(x, + d(y,$

Theorem 21:

Let (X, d) be a metric space. If $\langle x_n \rangle$ *and* $\langle y_n \rangle$ *are sequences in X, such that* $x_n \to x$, $y_n \to y$, *then show that*

$$d(x_n, y_n) \to d(x, y).$$

Proof :

Suppose $\langle x_n \rangle$ *and* $\langle y_n \rangle$ are sequences in a metric space (X, d).

Let $x_n \to x$, $y_n \to y$.

To prove that

$$d(x_n, y_n) \to d(x, y).$$

Let x, y, z $\in$ X be arbitrary.

Firstly we shall prove that

$$| d(x, z) - (y, z)| \leq d(x, y) \qquad \text{...(3)}$$

Now write the proof of Theorem 20.

$\because$ $\quad x_n \to x, y_n \to y.$

$\therefore$ $\quad \varepsilon > 0, \exists\, n_0, m_0, \in N$ s.t.

$\forall\, n \geq n_0 \Rightarrow d(x_n, x) < \varepsilon/2$

$\forall\, n \geq m_0 \Rightarrow d(y_n, y) < \varepsilon/2$

Take $\quad k_0 = \max. \{m_0, n_0\}$

Then $\quad d(x_n, x) < \varepsilon/2,\ d(y_n, y) < \varepsilon/2\ \forall\, n \geq k_0$

Now

$$\begin{aligned} & |d(x_n, y_n) - d(x, y)| \\ &= |d(x_n, y_n) - d(x_n, y) + d(x_n, y) - d(x, y)| \\ &\leq |d(x_n, y_n) - d(x_n, y)| + |d(x_n, y) - d(x, y)| \\ &= |d(y_n, x_n) - d(y, x_n)| + |d(x_n, y) - d(x, y)| \\ &\leq d(y_n, y) + d(x_n, x), \text{ according to (3)} \\ &< \varepsilon/2 + < \varepsilon/2 = \varepsilon \end{aligned}$$

$\therefore\ |d(x_n, y_n) - d(x, y)| < \varepsilon/2\ \forall\, n \geq k_0$

This $\quad \Rightarrow d(x_n, y_n) \to d(x, y)$

Theorem 22:

Let (X, d) be a metric space. Let $\langle x_n \rangle$ be a Cauchy sequence in X and $\langle y_n \rangle$ be a sequence in X s.t.

$$d(x_n, y_n) < \frac{1}{n}\ \forall\, n \in N.$$

Then show that

(i) $\langle y_n \rangle$ is a Cauchy sequence,

(ii) $\langle y_n \rangle$ converges to x iff $\langle x_n \rangle$ converges to x.

Proof :

Let $\langle x_n \rangle$ be a Cauchy sequence. Then given $\varepsilon > 0$.

$\exists\ n_0 \in N$ s.t. $m, n \geq n_0$

$\Rightarrow \quad d(x_n, x_m) < \varepsilon/3$ ...(1)

Choose $m_0 \in N$ s.t. $\dfrac{1}{m_0} < \dfrac{\varepsilon}{3}$.

Then $\quad n \geq m_0 \Rightarrow \dfrac{1}{n} \leq \dfrac{1}{m_0} < \dfrac{\varepsilon}{3}$.

or, $\quad d(x_n, y_n) < \dfrac{1}{n} \leq \dfrac{1}{m_0} < \dfrac{\varepsilon}{3}$.

or, $\quad d(x_n, y_n) \leq \dfrac{\varepsilon}{3}\ \forall\ n \geq m_0$ ...(2)

Choose $k_0 = \max\{n_0, m_0\}$. Then $\forall$ n, m $\geq k_0$, we have

$$d(y_n, y_m) \leq d(y_n, x_n) + d(x_n, x_m) + d(x_m, y_m)$$

$$< \frac{\varepsilon}{3} + \frac{\varepsilon}{3} + \frac{\varepsilon}{3} \text{ by (1) \& (2)}$$

or, $\quad d(y_n, y_m) < \varepsilon$,

Showing that $\langle y_n \rangle$ is a Cauchy sequence.

(ii) Let $\lim x_n = x$, then $\exists\ m_1 \in N$ s.t.

$d(x_n, x) < \varepsilon/3\ \forall\ n \geq m_1$ ...(3)

But $\quad d(y_n, x) \leq d(y_n, x_n) + d(x_n, x)$ ...(4)

Choose $\quad m_2 = \max\{m_1, m_0\}$

Using (2) and (3), in (4).

$$d(y_n, x) < \frac{\varepsilon}{3} + \frac{\varepsilon}{3} = \frac{2\varepsilon}{3} < \varepsilon\ \forall\ n \geq m_2$$

This $\Rightarrow \lim y_n = x$

Similarly, $y_n \to x \Rightarrow x_n \to x$.

Theorem 23:

Any sequence in a metric space converges to the unique point.

Or, A sequence in a metric space can converge to almost one point.

Or, If $\langle a_n \rangle$ is a sequence in a metric space (X, ρ) s.t. $a_n \to a_0$, $a_n \to a_0'$, then $a_0 = a_0'$.

Proof :

Let $\langle a_n : n \in N \rangle$ be a sequence in metric space (X, ρ) s.t.

$a_n \to a_0, a_n \to a_0'$,

To prove that $a_0 = a_0'$.

Suppose the contrary.

Then $a_0 \neq a_0'$.

Let $\rho(a_0, a_0') = 2r$.

Then $\quad S_r(a_0) \notin S_r(a_0') = \phi$.

$\because \quad a_n \to a_0, a_n \to a_0'$

$\therefore$ Given any $r > 0$, $\exists\ n_0, n_0' \in N$ s.t.

$$\forall\ n \geq n_0 \Rightarrow a_n \in S_r(a_0)$$
$$\forall\ n \geq n_0' \Rightarrow a_n \in S_r(a_0')$$

Taking $k_0 \Rightarrow \max\{n_0, n_0'\}$, we obtain

$$\forall\ n \geq k_0 \Rightarrow a_n \in S_r(a_0)\ a_n \in S_r(a_0')$$
$$\Rightarrow a_n \in S_r(a_0) \cap S_r(a_0')$$
$$\Rightarrow S_r(a_0) \cap S_r(a_0') \neq \phi.$$ A contradiction.

Hence our initial assumption $a_0 \neq a_0'$ is wrong. Consequently

$$a_0 = a_0'.$$

Theorem 24:

Let $\langle x_n \rangle$ be a sequence in a metric space (X, d) and the sets A_n be defined by

$$A_1 = \{x_1, x_2, x_3, \ldots\},$$
$$A_2 = \{x_2, x_3, \ldots\},$$
$$A_n = \{x_n, x_{n+1}, \ldots\},\ \forall\ n \in N.$$

Then the sequence $< x_n >$ is a Cauchy sequence iff $d(A_n) = 0$ as $n \to \infty$.

Proof :

Let $\langle x_n \rangle$ be a sequence in a metric space (X, d) and let

$$A_n = \{x_n, x_{n+1}, \ldots\},\ \forall\ n \in N \text{ so that}$$

and
$$\left.\begin{array}{r} n < m \Rightarrow A_m \subset A_n \\ r \geq n \Rightarrow x_r \in A_n \end{array}\right\} \quad \ldots(1)$$

Step (i) : Let $\langle x_n \rangle$ be a Cauchy sequence so that given any

$$\varepsilon > 0, \exists\ n_0 \in N \text{ s.t.}$$

$$n, m \geq n_0 \Rightarrow d(x_n, x_m) < \varepsilon \quad ...(2)$$

To prove that $d(A_n) = 0$ as $n \to \infty$.

Let $n \geq n_0$, then

$$d(A_n) = \sup\{d(x_r, x_s) : x_r, x_s \in A_n\}$$
$$= \sup\{d(x_r, x_s) : r, s \geq n\} < \varepsilon, \text{ by (1), (2)}$$

$\therefore \quad d(A_n) < \varepsilon \ \forall\ n \geq n_0$

This $\quad \Rightarrow \lim_{n \to \infty} d(A_n) = 0$

Step (ii) : Let us suppose $\lim_{n \to \infty} d(A_n) = 0$, so that given any $\varepsilon > 0$, $\exists\ n_0 \in N$ s.t.

$$n \geq n_0 \Rightarrow d(A_n) < \varepsilon$$

$\Rightarrow \quad d(A_{n0}) < \varepsilon.$

To prove that $< x_n >$ is a Cauchy sequence, it is enough to prove that

$$d(x_r, x_s) < \varepsilon \ \forall\ r, s \geq n_0.$$

Now, $r, s \geq n_0 \Rightarrow x_r, x_s \in A_{n0}$, by (1)

$$\Rightarrow d(x_r, x_s) < \varepsilon$$

$\because \quad d(A_{n0}) < \varepsilon$

Theorem 25:

Let A be a subset of a metric space (X, d).

Then A is non-dense in X $\Leftrightarrow (\overline{A}) = X - \overline{A}$ is dense in X.

Or

A closed set is nowhere dense $\Leftrightarrow$ its complement is everywhere dense.

Proof :

Recall that a is dense in X if $\overline{A} = x$ and A is nowhere dense in X if int $(\overline{A}) = \phi$.

A is non-dense (or no where dense) in X

$\Leftrightarrow$ int $(\overline{A}) = \phi \Leftrightarrow \exists$ no open nbd N of any point of $\overline{A}$ s.t. $N \subset \overline{A}$

$\Leftrightarrow \overline{A}$ contains no nbd. $\Leftrightarrow S_r(x) \not\subset \overline{A}\ \forall\ r > 0\ \&\ \forall\ x \in X$

$\Leftrightarrow S_r(x) \cap (\overline{A})' \neq \phi$

$\Leftrightarrow$ x is adherent point of $(\overline{A})'\ \forall\ x \in X$

$\Leftrightarrow [(\overline{A})'] = X = \Leftrightarrow (\overline{A})'$ is dense in X.

Theorem 26:

Let A and B be disjoint closed sets in a metric space (X, d). Then $\exists$ disjoint open sets G, H s.t. $A \subset G$, $B \subset H$.

Proof:

If either A or B is empty, say $A = \phi$, then $\exists$ disjoint open sets ϕ, X s.t. $A \subset \phi$, $B \subset X$.

Hence we may assume $A \neq \phi$, $B \neq \phi$.

Let $a \in A$, $b \in B$ arbitrary.

$$A \cap B = \phi,$$

$$\Rightarrow \quad a \notin B, b \notin A$$

$$\Rightarrow \quad d(a, B) > 0, d(b, A) > 0.$$

Take $d(a, B) = \delta_a > 0$ and $d(b, B)\ \delta_b > 0$.

Set $\quad S_a = S(a, 1/3\ \delta_a), s_b = (b, 1/3\ \delta_b)$.

Evidently $\quad a \in S_a, b \in S_b$.

Write $\quad G = \cap \{S_a : a \in A\}, H = \cap \{S_b : a \in B\}$

Being arbitrary union of open spheres, G and H are open sets.

$a \in S_a \Rightarrow A \subset G$ and $b \in S_b \Rightarrow B \subset H$.

Now we claim $G \cap H = \phi$.

If possible let $p \in G \cap H$, then $p \in G$, $p \in H$.

Consequently $\exists\ a_0 \in A$, $b_0 \in B$ s.t.

$p \in S(a_0, 1/3\ \delta_{a0})$, $p \in S(b_0, 1/3\ \delta_{b0})$. Let $d(a_0, b_0) = \varepsilon > 0$.

Then $d(a_0, B) = \delta_{a0} \leq \varepsilon$, $d(b_0, A) = \delta_{b0} \leq \varepsilon$

$$\varepsilon = d(a_0, b_0) \leq d(a_0, p) + d(p, b_0)$$

$$< \frac{1}{3}\delta_{a0} + \frac{1}{3}\delta_{b0} < \frac{\varepsilon}{3} + \frac{3}{3} = \frac{2\varepsilon}{3} \text{ or } \varepsilon < \frac{2\varepsilon}{3}.$$

which is not possible. Hence $G \cap H = \phi$.

Theorem 27:

Let (X, d) be a metric space. Suppose A is a closed subset of X and x is a point of X s.t. $x \notin A$. Then $\exists$ disjoint open subsets U and V of X s.t. $x \in U$, $A \subset V$.

Proof:

$x \notin A \Rightarrow \exists\ \delta > 0$ s.t. $S_\delta(x) \cap A = \phi$.

Choose ε, η with $0 < \varepsilon < \eta < \delta$.

write $U = S_\varepsilon(x)$, $V = X - S_\eta[x]$.

Since each open sphere is an open set and each closed sphere is a closed set. Also complement of closed set is open set. These prove that U and V are open sets s.t. $U \cap V = \phi$.

For $U = S_\varepsilon(x) \subset S_\eta(x) \subset S_\delta(x)$.

Also $A \subset V$, by construction of V.

Theorem 28:

Let $<f_n(x)>$ be a sequence of continuous functions defined on a metric space (X,d). Let this sequence converge uniformly to f on X. Then f(x) is continuous on X.

Or

Uniform limit of a sequence of continuous function is continuous.

Proof:

Since $<f_n(x)>$ converges uniformly to f on (X, d). Hence given $\varepsilon > 0$, $\exists\, n_0 \in N$ independent of $x \in X$ s.t. $n \geq n_0$.

$$\Rightarrow d(f_n(x), f(x) < \varepsilon/3 \quad ...(1)$$

Let $a \in X$ be arbitrary. To prove that f is continuous on X, we have to prove that f is continuous at x = a. For this we have to show that given $\varepsilon > 0$, $\exists\, \delta > 0$, 0 s.t. $d(x, a) < \delta$.

$$\Rightarrow d(f(x), f(a) < \varepsilon \quad ...(2)$$

Continuity of f_n at $a \in X$

$$\Rightarrow d(f_n(x), f(x) < \varepsilon/3 \text{ for } d(x, a), \delta \quad ...(3)$$

By (1), $d(f_n(a), f(a)) < \varepsilon/3 \ \forall\, n \geq n_0$ ...(4)

If $d(x, a) < \delta$, then

$d(f(x), f(a)) \leq d[f(x), f_n(x)] + d[f_n(x), f_n(a) + d[f_n(a), f(a)]$

$$< \frac{\varepsilon}{3} + \frac{3}{3} = \frac{\varepsilon}{3} o = \varepsilon \text{ by (1), (3), \& (4)}$$

or, $d(f(x), f(a)) < \varepsilon$ for $d(x, a) < \delta$. Hence the result (2).

Theorem 29:

Every non empty open set $G \subset R$ is the union of a countable collection of pairwise disjoint open intervals.

Proof:

Let G be a non-empty open subset of R and $x \in G$ be arbitrary. Then $\exists$ an interval I_x with centre x s.t. $x \in G$.

Let C_x denote the union of all such open intervals then $C_x = \cup \{I_x : x \in I_x \subset G\}$. Evidently C_x is the largest open interval containing x s.t. $x \in C_x \subset G$.

Moreover this also suggests that $G = \cup C_x$.

$\forall$ x, y $\in C_y$ we have either $C_x \cap C_y = \phi$ or $C_x \cap C_y \neq \phi$

Consider the case $C_x \cap C_y \neq \phi$.

Evidently $C_x \cap C_y$ is an open interval containing both x and y s.t. $C_x \subset C_x \cup C_y \subset G$,

$$C_y \subset C_x \cup C_y \subset G.$$

By def. of C_x, and C_y, $C_x \cup C_y \subset C_x$,

$$C_x \cup C_y \subset C_y$$

Hence $\quad C_x = C_x \cup C_y = C_y$

or $\quad C_x = C_y$

$\therefore \forall$ x, y $\in$ G we have either $C_x \cap C_y = \phi$ or $C_x = C_y$

s.t. $\quad G = \cup C_x$.

Second Part: It remains to show that the intervals C_x form a countable collection. Let $G = \{C_x\}$ be a collection of open intervals s.t. $C_n \neq \Phi \ \forall$ n and $C_n \cap C_m = \phi$ for $n \neq m$.

To prove that G is countable.

We know that each C_n contains an infinite number of rationals. Let K be the set of all rationals belonging to $\cup C_n$. Define a map $f : K \to G$ s.t.

f (x) = C_x where $C_x \in G$ s.t. $x \in C_x$.

For each $x \in K$, there is unique open interval C_x containing x, the rational number. Hence f is a well defined map. For each $C_a \in G$, has at least one rational $a \in K$ s.t. f(a) = C_a, showing thereby f is onto.

Also $\quad f(x) = f(y) \Rightarrow C_x = C_y$

where $\quad x \in C_x, y \in C_y \Rightarrow x = y$

Now f is one-one onto. Hence K ~ G. But K is countable. Hence G is countable.

For $K \subset Q$, Q is countable $\Rightarrow$ K is countable.

Theorem 30:

In a metric space (X, d), prove that

F is closed $\Leftrightarrow D(F) \subset F$.

Or

Prove that a subset F of a metric space X contains all its limit points iff X – F is open.

Proof :

Let (X, d) be a metric space and $F \subset X$.

We know that F is closed $\Leftrightarrow$ X – F is open.

Step I : Let X – F be open so that F is closed.

Aim. $D(F) \subset F$.

Let $x \in X - F$ be arbitrary. The X – F is an open set containing x s.t. $(X - F) \cap F = \phi$.

This $\Rightarrow$ is not a limit point of F

$$\Rightarrow x \notin D(F) \Rightarrow x \in X - D(F)$$

Thus, " $x \in X - F \Rightarrow x \in X - D(F)$

$\therefore$ $X - F \subset X' - D(F)$

or, $D(F) \subset F$

Step II : Given $D(F) \subset F$ ...(1)

To prove F is closed.

Let $y \in X - F$, then $y \notin F$

$$y \notin F,\ D(F) \subset F \Rightarrow y \notin D(F)$$

$\Rightarrow \exists$ open set G with $y \in G$ s.t.

$(G - \{y\}) \cap F = \phi$

$\Rightarrow G \cap F = \phi$ as $y \notin F$

$\Rightarrow G \subset X - F$

Thus we have shown that

any $y \in X - F \Rightarrow \exists$ open set G with $y \in G$ s.t. $G \subset X - F$

This $\Rightarrow$ X – F is open $\Rightarrow$ is closed.

Theorem 31:

Frechet Space: *Let F be the set of infinite sequences of real numbers.*

x, y, z $\in F$, *then*

$$x = \langle x_n \rangle = \langle x_1, x_2 \ldots \rangle,$$

$$y = \langle y_n \rangle,\ z = \langle z_n \rangle$$

where $x_n, y_n, z_n \in R$

Proof:

We define a map

$$d : F \times F : \rightarrow R \text{ s.t.}$$

$$d(x,y) = \sum_{n=1}^{\infty} \frac{1}{2^n} \frac{(x_n - y_n)}{\left[1 + \left|x_{n} - y_n\right|\right]}$$

To show that d is metric on F.

(i) $d(x, y) \geq 0$ For $|x_n - y_n| \geq 0 \ \forall_n$

(ii) $d(x, y) = 0 \Leftrightarrow x = y$

For $d(x,y) = 0 \Leftrightarrow \sum_{n=1}^{\infty} \frac{1}{2^n} \frac{(x_n - y_n)}{\left[1 + \left|x_{n} - y_n\right|\right]} = 0$

$$\Leftrightarrow \frac{1}{2^n} \frac{(x_n - y_n)}{\left[1 + \left|x_{n} - y_n\right|\right]} = 0 \ \forall \ n$$

$$\Leftrightarrow |x_n + y_n| = 0 \ \forall \ n$$

$$\Leftrightarrow |x_n + y_n| = \forall \ n$$

$$\Leftrightarrow x = y$$

(iii) $d(x, y) = d(y, x)$

For $|x_n + y_n| = |y_n + x_n|$

(iv) $d(x, y) \geq d(x, z) + d(z, y)$

Here we use the fact that

$$\frac{|\alpha + \beta|}{1 + |\alpha + \beta|} \leq \frac{|\alpha|}{1 + |\alpha|} + \frac{|\beta|}{1 + |\beta|}$$

In view of this we have

$$d(x,y) = \sum_{n=1}^{\infty} \frac{|x_n - y_n|}{1 + |x_{n} - y_n|} \cdot \frac{1}{2^n}$$

$$= \sum_{n=1}^{\infty} \frac{1}{2^n} \cdot \frac{|(x_n - z_n) + (z_n - y_n)|}{1 + |(x_n - z_n) + (z_n - y_n)|}$$

$$\geq \sum_{n=1}^{\infty} \frac{1}{2^n} \cdot \frac{|x_n - z_n|}{1 + |x_n - z_n|} + \sum_{n=1}^{\infty} \frac{1}{2^n} \cdot \frac{|z_n - y_n|}{1 + |z_n - y_n|}$$

$= d(x, z) + d(z, y).$

Thus d is metric on F. The pair (F, d) is a metric space and this metric space is called *Frechet space.*

SOLVED EXAMPLES

Example 1: *In any metric space, show that*

$$x - \overline{A} = (X - A)^o$$

or, $(\overline{A})' = (A')^o$

Solution: $(\overline{A})' = X - \overline{A}$

= X – Intersection of all closed super sets of A.

$= X - \bigcap_i F_i$ where F_i is closed and $F_i \supset A$.

$= \bigcup_i (X, F_i)$ where $X - F_i$ is open

and $X - F_i \subset X - A$

= Union of open subsets of X – A = A'

$= (A')^c$

Example 2: *Given an example of Pseudo metric which is not a metric.*

Solution: Suppose C is the family of all Riemann integral functions f; $[0, 1] \to R$.

Let f, g, h, $\in$ C be arbitrary. Define a map

$d : C \times C \to R$ s.t. $d(f, g) = \int_0^1 |f - g|(x)\, dx$.

To prove d is a Pseudo metric but; not metric on C.

$d(f, g) \geq$ For $|f - g|(x) \geq 0$ and so.

$\int_0^1 |f - g|(x)\, dx \geq 0$

(1) d (f, g) = d (g, f).

This follows form the fact that $|f - g| = |g - f|$

(2) d (f, g) + d (g, f).

$= \int_0^1 |f - g|(x)\, dx + \int_0^1 |g - h|(x)\, dx$

$= \int_0^1 [|f - g| + |g - h|](x)\, dx$

$\geq \int_0^1 |f - g + g - h|(x)\, dx$

$= \int_0^1 |f - h|(x)\, dx = d(f, h)$

(3) d (F, g) = 0 if f = g

If f = g, then

$$d(f, g) = \int_0^1 |f - g|(x)\, dx = \int_0^1 |g - g|(x)\, dx = \int_0^1 0\, dx = 0$$

(4) d (f, g) = 0 does not necessarily imply f = g.

That is to say, d (f, g) = 0

$\Rightarrow$ f $\neq$ g, in general.

Solution:

Suppose $f(x) = \begin{cases} 4\, where\, x = 0 \\ 0\, when\, 0 < x \le 1 \end{cases}$

and $g(x) = \begin{cases} 9\, when\, x = 0 \\ 0\, when\, 0 < x \le 1 \end{cases}$

The f(x) $\neq$ g (x), so that f $\neq$ g.

Also $|f - g|(x) = \begin{cases} 9 - 4 = & when\, x = 0 \\ 0 & when\, 0 < x \le 1 \end{cases}$

Evidently f – g is Riemann integrable,

x = o is the only point of discontinuity.

$$d(f, g) = \int_0^1 |f - g|(x)\, dx = \int_0^1 0\, dx = 0$$

Thus, d (f, g) = 0 but f $\neq$ g.

This proves that d is a Pseudo metric but metric on C.

Similar Ex : *Let C be the class of bounded continuous functions f : [0, 1 $\rightarrow$ R. Define a map d : C × C $\rightarrow$ R such that*

$$d(f, g) = \int_0^1 |f - g|(x)\, dx\ \forall f, g \in C = \|f - g\|$$

Then d is a metric on C.

Example 3: *Show that C the set of complex numbers is metric space under the metric d defined as follows:*

$$d(z_1, z_2) = \frac{|z_1 - z_2|}{\left[1 + |z_1|^2\right]^{1/2} \left[1 + |z_2|^2\right]^{1/2}}$$

Solution: Let z_1, z_2, z_3 be arbitrary complex numbers . It is given that

$$d(z_1, z_2) = \frac{|z_1 - z_2|}{\left[1 + |z_1|^2\right]^{1/2} \left[1 + |z_2|^2\right]^{1/2}}$$

To prove that (C, d) is a metric space.

We know that |z| $\geq$ 0 for every complex number z.

If view of this, (1) says that $d(z_1, z_2) \geq 0$. ...(1)

Since $|z_1 - z_2| = |z_2 - z_1|$. In view of this, (1) $\Rightarrow$

$$d(z_1, z_2) = d(z_2, z_1) \qquad ...(2)$$

$$d(z_1, z_2) = 0 \Leftrightarrow \frac{|z_1 - z_2|}{\left[1+|z_1|^2\right]^{1/2}\left[1+|z_2|^2\right]^{1/2}} = 0$$

$$\Leftrightarrow |z_1 - z_2| = 0 \Leftrightarrow z_1 = z_2$$

$$\therefore \quad d(z_1, z_2) = 0 \Leftrightarrow z_1 = z_2 \qquad ...(3)$$

$d(z_1, z_2) + d(z_2, z_3)$

$$= \frac{|z_1 - z_2|}{\left[1+|z_1|^2\right]^{1/2}\left[1+|z_2|^2\right]^{1/2}} + \frac{|z_2 - z_3|}{\left[1+|z_2|^2\right]^{1/2}\left[1+|z_3|^2\right]^{1/2}}$$

$$\geq \frac{|z_1 - z_2|}{\left[1+|z_1|^2\right]^{1/2}\left[1+|z_2|^2\right]^{1/2}\left[1+|z_3|^2\right]^{1/2}}$$

$$+ \frac{|z_2 - z_3|}{\left[1+|z_1|^2\right]^{1/2}\left[1+|z_2|^2\right]^{1/2}\left[1+|z_3|^2\right]^{1/2}}$$

$$= \frac{|z_1 - z_2| + |z_2 - z_3|}{\left[1+|z_1|^2\right]^{1/2}\left[1+|z_2|^2\right]^{1/2}\left[1+|z_3|^2\right]^{1/2}}$$

$$\geq \frac{|z_1 - z_3|}{\left[1+|z_1|^2\right]^{1/2}\left[1+|z_2|^2\right]^{1/2}\left[1+|z_3|^2\right]^{1/2}}$$

$$\geq \frac{|z_1 - z_3|}{\left[1+|z_1|^2\right]^{1/2}\left[1+|z_3|^2\right]^{1/2}} = d(z_1, z_3)$$

or $\quad d(z_1, z_2) + d(z_2, z_3) \geq d(z_1, z_3)$ ...(4)

From (1), (2), (3) and (4), it follows that (C, d) is a metric space.

Example 4: *Show that the function d defined by d (p, d) = $\|p - q\|$, where p, q are vectors in a norma! space V, is a metric on V.*

Solution: If p, q, r $\in$ V and K is any scalar, then by def. of normed vector space,

(N_1) $\|p - q\| \geq 0$, (N_2) $\|p - q\| = 0 \Leftrightarrow p = 0$

(N_3) $\|kp\| = |k|\ \|p\|$, (N_4) $\|p+q\| \le \|p\| + \|q\|$

We have $d(p, q) = \|p - q\|$ ...(1)

Using (N_1) in (1), we find that $d(p, q) \ge 0$ ...(M_1)

Using (N_2) in (1), we see that

$$d(p, q) = 0 \Leftrightarrow \|p-q\| = 0 \Leftrightarrow p = q = 0$$

$$\Leftrightarrow p = q \quad ...(M_2)$$

or $\quad d(p, q) = 0 \Leftrightarrow p = q$ ...(M_3)

Using (N_3) in (1), we see that

$$d(q,p) = \|q-p\| = \|(-1)(p-q)\| = \|p-q\| = d(p,q)$$

or $\quad d(q, p) = d(p, q)$ ...(M_3)

Using (N_4) in (1), we see that

$$d(p,r) = \|p-r\| = \|(p-q) + (q,r)\| \le \|p-q\| + \|q-r\|$$

$$d(p, q) + d(q, r)$$

or $\quad d(p, r) \le d(p, q) + d(q, r)$

From (M_1), (M_2), (M_3) and (M_4), it follows that (C, d) is a metric space.

Example 5: *In a metric space (X, d), $S = S(x_0, r)$ is an open sphere and p is any point of it. Prove that $\exists$ an open sphere $T = S(p, \varepsilon)$ s.t. $T \subset S$.*

Solution: $p \in S(x_0, r) \Rightarrow d(p, x_0) < r \Rightarrow r - d(p, x_0) = r_1 > 0$.

Take $\varepsilon < r_1$, then $x \in S\ p\ \varepsilon \Rightarrow d(x, p) < \varepsilon < r_1 \Rightarrow d(x, p) < r_1$

or $\quad d(x, p) < r - d(p, x_0)$

or $\quad d(x, p) + d(p, x_0) < r$

or $\quad d(x, x_0) \le d(x, p) + d(p, x_0) < r$

or $\quad d(p, x_0) < r$

or $\quad x \in S(x_0, r)$ ...(M_4)

Thus $\quad x \in S(p, \varepsilon) \Rightarrow x \in S(x_0, r)$.

Hence $\quad S(p, \varepsilon) \subset S(x_0, r)$ or $T \subset S$.

Example 6: *If G_1, G_2 are open sets in metric space (X_1, d_1) and (X_2, d_2) then $G_1 \times G_2$ is open in the product metric space (X, d).*

Solution: Let $x = (x_1, x_2)$, $y\ (y_1, y_2) \in X = X_1 \times X_2$, then $x_1, y_1 \in X_1$ and $x_2\ y_2 \in X_2$. Let $x \in G_1 \times G_2$, then $x_1 \in G_1$, $x_2 \in G_2$.

By def. of open set,

$x \in G_1 \Rightarrow \exists\, r_1 > 0$ s.t. $S(x_1, r_1) \subset G_1$ ...(1)

Similarly, $S(x_2, r_2) \subset G_2$ ...(2)

By (1) and (2),

$S(x_1, r_1) \times S(x_2, r_2) \subset G_1 \times G_1$ ...(3)

Take $r = \min\{r_1, r_2\}$, then $r \le r_1$, $r \le r_2$

$\therefore\quad S(x_1, r) \times S(x_2, r) \subset S(x_1, r_1) \times S(x_2, r_2) \subset G_1 \times G_2$

or $\quad S(x, r) \subset G_1 \times G_2$, for any $x \in G_1 \times G_2$.

By def., this proves that $G_1 \times G_2$ is open.

Example 7: *In a metric space, prove that $d(A) = d(\overline{A})$.*

Solution: Evidently $A \subset B \Rightarrow d(A) \le d(B)$.

But $A \subset \overline{A}$ and so that $d(A) \le d(\overline{A})$. ...(1)

Let $x, y \in \overline{A}$, then \$ u, v $\in$ A s.t.

$$d(x, u) < r,\ d(u, v) < r \text{ where } r > 0.$$

But $d(x, y) \le d(x, u) + d(u, v) + d(v, y)$

$$< d(u, v) + 2r \le d(A) + 2r$$

or $\qquad d(x, y) \le d(A) + 2r.$

For $d(A) = \sup d(x, y)$ where $x, y \in A$.

This $\qquad \Rightarrow d(\overline{A}) \le d(A)$. ...(2)

Combining (1) and (2), $d(A) = d(\overline{A})$.

Example 8: *Let X be a non-empty set and let d be a real valued function of ordered pairs of elements of X which satisfies the following conditions:*

(i) $d(x, y) = 0 \Leftrightarrow x = y$

(ii) $d(x, y) \le d(x, z) + d(y, z)$

Then prove that d is a metric on X.

Solution:

(1) To prove $d(x, y) \ge 0$.

By (ii), $d(x, x) \le d(x, z) + d(x, z)$, on putting $x = y$.

or, $0 \le 2d(x, z)$ by (i)

or, $d(x, z) \ge 0$

or, $d(x, y) \ge 0$

(2) To prove $d(x, y) = d(y, x)$.

putting z = x in (ii),

$$d(x, y) \leq 0 + d(y, x)$$

or, $d(x, y) \leq d(y, x)$

Interchanging x and y,

$d(y, x) \leq d(x, y)$

Combining the last two, $d(x, y) = d(y, x)$.

(3) $d(x, y) = 0 \Leftrightarrow x = y$ by (i)

(4) To prove $d(x, z) + d(z, y) \geq d(x, y)$

Using (2) in (ii), $d(x, y) \leq d(x, z) + d(z, y)$

or, $d(x, z) + d(z, y) \geq d(x, y)$.

Example 9: *Every set in a discrete metric space is open.*

Solution: Let (X,d) be a discrete metric space. Let $x, y \in X$ be arbitrary.

By def. of discrete metric, $d(x, y) = \begin{cases} 1 \text{ if } x \neq y \\ 0 \text{ if } x = y \end{cases}$.

Let r be any positive real number s.t. $r \leq 1$.

Then $Sr(x) = \{y \in X : d(y, x) < r \leq 1\}$

$= \{y \in X : d(y, x) < 1\}$

$= \{y \in X : d(y, x) = 0$, by def. of d.

$= \{y \in X : y = x\} = \{0\}$.

or, $S_{r(x)} = \{0\}$.

But every open sphere is an open set.

$\therefore$ $\{x\}$ is an open set in X $\forall$ $x \in X$.

If $A = \{x_1, x_2, ..., x_n\}$ = finite set $\subset X$, then

$A = \bigcup_{r=1}^{n} \{x_r\}$ = finite union of open sets.

= open set.

Hence every finite sub set of X is open set. ...(1)

If $B = \{x_1, x_2, x_3, ...\} \subset X$, then

B is an infinite subset of X.

Now $B = \bigcup_{r=1}^{\infty} \{x_r\}$

= Arbitrary union of open sets

$= \text{open set, by Theorem 8.}$

$\therefore$ B is an open set, ...(2)

From (1) and (2), it follows that every subset (finite or infinite) is an open set in X.

Example 10: *Let ρ_1 and ρ_2 be metrics on the same set X, show that there exists a metric ρ on X such that*

$$\rho\,(x, y) \geq \rho_i\,(x, y) \text{ for } i = 1, 2$$

and $\forall\, x, y \in X$.

Solution: Let ρ_1 and ρ_2 bc metrics on the same set X, Let x, y, z $\in$ X be arbitrary.

To prove that $\exists$ a metric ρ on X s.t.

$$\rho\,(x, y) \geq \rho_i\,(x, y) \text{ for } i = 1, 2$$

Write $\quad \rho\,(x, y) = \rho_1\,(x, y) + \rho_2\,(x, y).$ ...(1)

We claim ρ is metric on X.

(i) $\quad \rho\,(x, y) \geq 0$

$\because \quad \rho_1\,(x, y) \geq 0\,,\ \rho_2\,(x, y) \geq 0.$

(ii) $\quad \rho\,(x, y) = 0$ iff $x = y$

For $\quad \rho\,(x, y) = 0 \Leftrightarrow \rho_1\,(x, y) + \rho_2\,(x, y) = 0.$

$$\Leftrightarrow \rho_1\,(x, y) = 0 = \rho_2\,(x, y).$$

Since ρ_1 and ρ_2 both are non-negative.

$$\Leftrightarrow x = y$$

(iii) $\quad \rho\,(y, x) = \rho\,(x, y)$

For $\rho\,(y, x) = \rho_1\,(y, x) + \rho_2\,(y, x) = \rho_1\,(x, y) + \rho_2\,(x, y)$

$$\rho\,(x, y)$$

$\because \quad \rho_1\,(x, y) = \rho_1\,(y, x),\ \rho_2\,(x, y) = \rho_2\,(y, x)$

(iv) $\rho\,(x, y) + \rho\,(y, z) \geq \rho\,(x, z)$

For $\rho\,(x, y) + \rho\,(y, z) = \rho_1\,(x, y) + \rho_2\,(x, y) + \rho_1\,(y, z) + \rho_2\,(y, z)$

$$[\rho_1\,(x, y) + \rho_1\,(y, z)] + [\rho_2\,(x, y) + \rho_1\,(y, z)]$$

$$^3\,\rho_1\,(x, z) + \rho_2\,(x, z) = \rho\,(x, z).$$

From what has been done it follows that ρ is a metric on X.

Moreover (1) $\Rightarrow \rho\,(x, y) \geq \rho_1\,(x, y)$ for i = 1, 2.

Hence ρ is the required metric.

Example 11: *Show that the map $d : R^2 \times R^2 \to R$*
define by $d(x, y) = \max. \{|x_1 - y_1|, |x_2 - y_2|\}$.
where $x = (x_1, x_2), y = (y_1, y_2) \in R_2$
is a metric on R^2.

Solution: Let $x = (x_1, x_2)$, $y = (y_1, y_2)$, $z = (z_1, z_2)$ be arbitrary elements or R^2. Also let

$d(x, y) = \max\{|x_1 - y_1|, |x_2 - y_2|\}$

(i) $d(x, y) \geq 0$.

Since modulus of any quantity ;is non-negative.

(ii) $d(x, y) = 0$ iff $x = y$.

For $d(x, y) = 0 \Leftrightarrow \max.\{|x_1 - y_1|, |x_2 - y_2|\} = 0$

$\Leftrightarrow |x_1 - y_1| = 0 = |x_2 - y_2|$

$\Leftrightarrow x_1 = y_1, |x_2 = y_2$

$\Leftrightarrow (x_1, x_2) = y_2, y_2)$

$\Leftrightarrow x = y$

(iii) $d(x, y) = d(y, x)$.

For $|a - b| = |b - a|$ and so

$\max.\{|x_1 - y_1|, |x_2 - y_2|\} = \max\{|y_1 - x_1|, |y_2 - x_2|\}$

(iv) $d(x, y) + d(y, z) \geq d(x, z)$

For $d(x, y) = \max\{|x_1 - z_1|, |x_2 - z_2|\}$

$= \max.\{|x_1 - y_1 + y_1 - z_1|, |x_2 - y_2 + y_2 - z_2|\}$

$\leq \max.\{|x_1 - y_1| + |y_1 - z_1|, |x_2 - y_2| + |y_2 - z_2|\}$

$\leq \max.\{|x_1 - y_1|, |x_2 - y_2|\} + \max.\{|y_1 - z_1|, |y_2 - z_2|\}$.

$= d(x, y) + d(y, z)$.

The above facts lead to the conclusion that d is a metric on R^2.

Example 12: *In every metric space, all finite sets are closed*

Solution: Let (X, d) be a metric space.

To prove that every finite subset of X is closed.

For any subset A of X, we know that

$\overline{A} = \{x \in X : d(x, A) = 0\}$

Let $p \in X$ be arbitrary.

Then $\{\overline{p}\} = \{x \in X : d(x, \{p\}) = 0)$

$$= \{p\}$$

$$\therefore \quad d(x, \{p\}) = 0 \Rightarrow x \in \{p\} \Rightarrow x = \dot{p}$$

$$\{p\} = \{\bar{p}\},$$

Showing thereby, $\{p\}$ is closed in X.

Hence every singleton set in X is closed. Moreover a finite union of closed sets is closed. A finite set is a finite union of singleton sets in which every singleton set is closed.

Example 13: *Let d be a Pseudo metric on X. Define a relation on X by setting X~y iff d (x, y) = 0. Show that this relation is an equivalence relation whose corresponding equivalence class can be made into a metric space in a natural way.*

Solution: Let d be a Pseudo metric on X so that

$\forall$ x, y, z, $\in$ X we have

(1) $d(x, y)\ ^3\ 0$

(2) $d(x, y) = d(y, x)$

(3) $d(x, x) = 0$

(4) $d(x, y) + d(y, z)\ ^3\ d(x, z)$

Define a relation ~ on X s.t.

$$x \sim y \text{ iff } d(x, y) = 0$$

To prove that ~ is an equivalence relation

(i) ~ is reflexive

For $\quad d(x, x) = 0$

$\Rightarrow \quad x \sim x$, by (3)

(ii) ~ is symmetric.

For $\quad x \sim y \Rightarrow d(x, y) = 0$

$\Rightarrow \quad d(y, x) = 0$, by (2)

$\Rightarrow y \sim x$

(iii) ~ is transitive.

For $\quad x \sim y, y \sim z$

$\Rightarrow \quad d(x, y) = 0 = d(y, z)$

$\Rightarrow \quad d(x, z) \le (x, y) + d(y, z) = 0 + 0 = 0$

$\Rightarrow \quad d(x, z) \le 0$

But $\quad d(x, z) \ge 0$, by (1)

This $\Rightarrow d(x, z) = 0 \Rightarrow x \sim z$

Thus we have proved that ~ is an equivalence relation of X.

Hence the relation ~ decomposes X into family of disjoint subsets of X, known as equivalence classes. Let [X] denote the equivalence class determined by $x \in X$ and let C be the set of all equivalence classes,

i.e., $C = \{[x] : x \in X\}$

Define a map $f : C \times C \rightarrow R$ s.t.

$$f([x_1], [x_2]) = d(x_1, x_2)$$

where $[x_1], [x_2] \in C$

Then f is a metric for C on account of the fact that d is a pseudometric on X. The metric f defined as above is called natural metric for C.

Example 14: *Let C be the set of all real valued bounded continuous functions defined on [0, 1].*

If d is defined by

$d(f, g0 = \sup \{|f(x) - g(x)| : x \in [0, 1]\}$

then d is a metric on C, where f, g∈ C.

Solution:

Let f, g, h ∈ C and x ∈ [0, 1] be arbitrary. Then

(1) $d(f, g) \geq 0$. For mod of any thing ≥ 0.

(2) $d(f, g) = d(g, f)$. For $|f(x) - g(x)| = |g(x) - f(x)|$.

(3) $d(f, g) = 0 \Leftrightarrow f = g$

For $\sup \{|f(x) - g(x)| : x \in [0, 1]\} = 0 \Leftrightarrow |f(x) - g(x)| = 0$

$$\Leftrightarrow f(x) - g(x) \quad \forall x$$

$$\Leftrightarrow f = g$$

(4) To prove $d(f, h) \leq d(f, g) + d(g, h)$

Since $|f - h| = |f - g + g - h|$

$$\leq |f - g| + |g - h|$$

This $\Rightarrow \sup |f(x) - h(x)| \leq \sup |f(x) - g(x)| + \sup |g(x) - h(x)|$

$$\Rightarrow d(f, h) \leq d(f, g) + d(g, h)$$

From what has been it follows that d is a metric on C.

Example 15: *Let X be a non-empty set and let d be a real valued function of ordered pairs of elements of X which satisfies the following conditions:*

(i) $d(x, y) = 0 \Leftrightarrow x = y$

(ii) $d(x, y) \leq d(x, z) + d(y, z)$

Then prove that d is a metric on X.

Solution:

(1) To prove $d(x, y) \geq 0$.

By (ii), $d(x, x) \leq d(x, z) + d(x, z)$, on putting $x = y$.

or, $0 \leq 2d(x, z)$ by (i)

or, $d(x, z) \geq 0$

or, $d(x, y) \geq 0$

(2) To prove $d(x, y) = d(y, x)$.

putting $z = x$ in (ii),

$$d(x, y) \leq 0 + d(y, x)$$

or, $d(x, y) \leq d(y, x)$

Interchanging x and y,

$d(y, x) \leq d(x, y)$

Combining the last two, $d(x, y) = d(y, x)$.

(3) $d(x, y) = 0 \Leftrightarrow x = y$ by (i)

(4) To prove $d(x, z) + d(z, y) \geq d(x, y)$

Using (2) in (ii), $d(x, y) \leq d(x, z) + d(z, y)$

or, $d(x, z) + d(z, y) \geq d(x, y)$

Remark : This gives an alternative definition of metric.

Example 16: *In any metric space (X, d), show that* $A \subset X$ *is closed iff* $A = \overline{A}$

Solution: Let (X, d) be metric space and $A \subset X$. Firstly we shall establish the following lemmas :

(i) $\overline{A}$ is closed and $A \subset \overline{A}$.
(ii) $F \supset A$, F is closed $\Rightarrow F \supset C$

By definition, $\overline{A} = \cap\{F \subset X : F \text{ is closed}, F \supset A\}$, ...(1)
(i) Since an arbitrary; intersection of closed sets is closed and hence $\overline{A}$ is closed.

By (1), $\overline{A} \supset A$
(ii) Suppose $F \subset X$ is closed and $F \supset A$.

$\therefore$ F is one of those members whose intersection is $\overline{A}$ and hence $F \supset \overline{A}$. .E.D. lemma (ii)

Now we come to the proof of actual theorem.

Let A = $\overline{A}$. We claim A is closed. By (i). $\overline{A}$ is closed.

$\overline{A}$ is closed, A = $\overline{A}$ $\Rightarrow$ is closed.

Conversely let A be closed. To prove that A = $\overline{A}$.

By the lemma (i), A $\subset$ $\overline{A}$. ...(2)

By the lemma (ii), f is closed, F $\supset$ A $\Rightarrow$ F $\supset$ $\overline{A}$. ...(3)

Since A $\supset$ A for every set A.

In particular (3) gives A is closed, A $\supset$ A $\Rightarrow$ A $\supset$ $\overline{A}$ ...(4)

From (2) and (4), A= $\overline{A}$.

This concludes the problem.

Example 17: *In* any metric space (X, d), prove that A is open

$\Leftrightarrow$ $A^o = A$.

Solution: Let A be a subset of a metric space (X, d). By def. of interior,

$$A^o = \cup \{S_{r(x)} : S_{r(x)} \subset A\} \quad ...(1)$$

Since every open sphere is an open set and arbitrary union of open sets is open. Consequently,

A^o is an open set. ...(2)

By (1), it is clear that $A^o \subset A$ ...(3)

and A^o is largest open subset of A. ...(4)

(i) Given $A = A^o$ ...(5)

Aim. A is an open set

(2) and (5) $\Rightarrow$ A is an open set.

(ii) Given A is an open set ...(6)

Aim, $A = A^o$

(4) and (6) $\Rightarrow$ = A A^o

Example 18: *Show that a subset F of a metric space X is closed iff* $\{x \in X : d(x, F) = 0\} \subset F$.

Solution: Let (X, d) be a metric space.

Let F $\subset$ X be closed.

To prove that

$\{x \in X : d(x, F) = 0\} \subset F.$

We know that $\overline{F} = \{x \in X : d(x, F) = 0\}.$

F is closed $\Rightarrow \overline{F} = F \Rightarrow \overline{F} \subset F$

$\Rightarrow \{x \in X : d(x, F) = 0\} \subset F.$ on using (1)

Conversely suppose that F is a subset of a metric space (X, d) with the property that

$$\{x \in X : d(x, F) = 0\} \subset F \quad ...(2)$$

To prove that F is closed, it suffices to show that $F = \overline{F}$.

But $\overline{F} = \{x \in X : d(x, F) = 0\}$ for any set $F \subset X$. ...(3)

Now (2) and (3) $\Rightarrow \overline{F} \subset F$ ...(4)

But $F \subset \overline{F}\ \forall F$ ...(5)

Combining (4) and (5), $\overline{F} = F$. Hence proved.

Example 19: *In every metric space, all finite sets are closed*

Solution: Let (X, d) be a metric space.

To prove that every finite subset of X is closed.

For any subset A of X, we know that

$\overline{A} = \{x \in X : d(x, A) = 0\}$

Let $p \in X$ be arbitrary.

Then $\{\overline{p}\} = \{x \in X : d(x, \{p\}) = 0)$

$= \{p\}$

$\therefore\quad d(x, \{p\}) = 0 \Rightarrow x \in \{p\} \Rightarrow x = p$

$\{p\} = \{\overline{p}\}$,

Showing thereby, {p} is closed in X.

Hence every singleton set in X is closed. Moreover a finite union of closed sets is closed. A finite set is a finite union of singleton sets in which every singleton set is closed. Accordingly every finite set in a metric space is closed.

Example 20: *If d is a metric on X, prove that the metric ρ defined by*

$$\rho(a,b) = \frac{d(a,b)}{1 + d(a,b)}$$

is also a metric on X.

Solution: Do yourself.

Example 21: *Let R be the set of all real numbers and let*

$$d(x,y) = \frac{|x-y|}{1+|x-y|} \text{ for all } x, y, \in R.$$

Prove that d is a metric for R.

Solution: Let x, y, z ∈ R be arbitrary.

Let $\rho(x,y) = |x-y|$.

Then (1) $\Rightarrow \rho(x, y) \geq 0$.

For modulus of any quantity is always non-negative.

(2) $\rho(x, y) = 0$ iff $x = y$.

For $\rho(x, y) = 0$

$$\Rightarrow |x-y| = 0 \Rightarrow x = y.$$

(3) $\rho(x, y) = \rho(y, x)$.

For $|x - y| = |y - x|$

(4) $\rho(x, y) + \rho(y, z) \geq \rho(x, z)$.

For $\rho(x, y) + \rho(y, z) = |x - y| + |y - z|$.

$$\geq |x - y + y - z| = |x - z| = \rho(x, z).$$

i.e., $\rho(x, y) + \rho(y, z) \geq \rho(x, z)$.

The above arguments prove that ρ is a metric on R.

By what is given $d(x, y) = \frac{|x-y|}{1+|x-y|}$.

This is expressible as

$$d(x, y) = \frac{\rho(x,y)}{1+\rho(x,y)}$$

where ρ is a metric on R. We have to prove that d is a metric on R.

Example 22: *Let ρ_1 and ρ_2 be metrics on the same set X, show that there exists a metric ρ on X such that*

$$\rho(x, y) \geq \rho_i(x, y) \text{ for } i = 1, 2$$

and $\forall x, y \in X$.

Solution: Let ρ_1 and ρ_2 be metrics on the same set X, Let x, y, z ∈ X be arbitrary.

To prove that ∃ a metric ρ on X s.t.

$\rho(x, y) \geq \rho_1(x, y)$ for $i = 1, 2$

Write $\rho(x, y) = \rho_1(x, y) + \rho_2(x, y)$. ...(1)

We claim ρ is metric on X.

(i) $\rho(x, y) \geq 0$

$\because$ $\rho_1(x, y) \geq 0$, $\rho_2(x, y) \geq 0$.

(ii) $\rho(x, y) = 0$ iff $x = y$

For $\rho(x, y) = 0 \Leftrightarrow \rho_1(x, y) + \rho_2(x, y) = 0$.

$\Leftrightarrow \rho_1(x, y) = 0 = \rho_2(x, y)$.

Since ρ_1 and ρ_2 both are non-negative.

$\Leftrightarrow x = y$

(iii) $\rho(y, x) = \rho(x, y)$

For $\rho(y, x) = \rho_1(y, x) + \rho_2(y, x) = \rho_1(x, y) + \rho_2(x, y)$

$\rho(x, y)$

$\because$ $\rho_1(x, y) = \rho_1(y, x), \rho_2(x, y) = \rho_2(y, x)$

(iv) $\rho(x, y) + \rho(y, z) \geq \rho(x, z)$

For $\rho(x, y) + \rho(y, z) = \rho_1(x, y) + \rho_2(x, y) + \rho_1(y, z) + \rho_2(y, z)$

$[\rho_1(x, y) + \rho_1(y, z)] + [\rho_2(x, y) + \rho_1(y, z)]$

${}^3\ \rho_1(x, z) + \rho_2(x, z) = \rho(x, z)$.

From what has been done it follows that ρ is a metric on X.

Moreover (1) $\Rightarrow \rho(x, y) \geq \rho_1(x, y)$ for $i = 1, 2$.

Hence ρ is the required metric.

Example 23: *Let l_∞ denote the set of all bounded sequences $\langle x_n \rangle$. If $x = \langle x_n \rangle$, $y = \langle y_n \rangle$ in $l_{\infty,}$ then we define*

$$d(x, y) = \sup \{|x_n - y_n| : n \in N\}.$$

Show that d is a metric on $l_{\infty.}$

Solution: (1) $d(x, y) = \sup |x_n - y_n|$

$\leq \sup |x_n| + \sup |y_n| < \infty$

or d (x, y) is finite. For $<x_n>$ and $<y_n>$ both are bounded.

This $\Rightarrow$ d (x, y) is well defined map.

(2) Evidently $d(x, y) \geq 0$.

(3) $d(x, y) = 0$ iff $x = y$

For $x = y \Leftrightarrow x_n = y_n \ \forall \ n \Leftrightarrow |x_n - y_n| = 0 \ \forall \ n$

$\Leftrightarrow \sup |x_n - y_n| = 0 \Leftrightarrow d(x, y) = 0$

(4) $d(x, y) = d(y, x)$.

For $|x_n - y_n| = |y_n - x_n|$

(5) $d(x, y) \leq d(x, z) + d(z, y)$

For $d(x, y) = \sup |x_n - y_n| = \sup |x_n - z_n + z_n - y_n|$

$\leq \sup [|x_n - z_n| + |z_n - y_n|]$

$\leq \sup |x_n - z_n| + \sup |z_n - y_n|$

$= d(x, z) + d(z, y)$

$\therefore$ (l_∞, d) is a metric space.

Example 24: *Every derived set in a metric space is a closed set.*

Solution: Let (X, d) be a metric space and $A \subset X$. Let x be a limit point of D (A) . If we show that x ε D (A), the result will be proved.

x is limit point of D (A) $\Rightarrow (S_r(x) - \{x\} \cap D(A) \neq \varnothing$ ...(1)

for every open sphere $S_r(x)$.

Then $\exists \ y_r \neq x \in S_r(x) \cap D(A)$

i.e. $y_r \in D(A)$, $y_r \in S_r(x)$

$\therefore$ $d(x, y) < r$.

Take $r - d(x, y) = r_1 > 0$.

Evidently $S_{r1}(y_r) \subset S_r(x)$ ...(2)

$$\left[\begin{array}{l} For\ z \in S_{r1}(y_r) \Rightarrow d(z, y_r) < r_1 = r - d(x, y_r) \\ \qquad \Rightarrow d(x, y_r) + d(y_r, z) < r \\ \qquad \Rightarrow d(x, z) < r \Rightarrow z \in S_r(x) \end{array}\right]$$

$y_r \in D(A) \Rightarrow S_{r1}(y_r) \cap A \neq \varnothing \ \forall \ S_{r1}(y_r)$

$\Rightarrow$ $S_r(x) \cap A \neq \varnothing$, by (2)

$\Rightarrow$ $\in D(A)$.

Example 25: *Prove that the mapping defined by $d : R \times R \to R$ such that*

$$d(x,y) = \begin{cases} o \ if \ x = y \\ 1 \ if \ x \neq y \end{cases}$$

is a metric on R.

Solution: Let $x, y, z \in R$ be arbitrary and let

$$d(x,y) = \begin{cases} o \text{ if } x = y \\ 1 \text{ if } x \neq y \end{cases}$$

Now (i) d (x, y) ≥ 0

For d (x, y) = 0 or 1.

(ii) d (x, y) = 0 or iff x = y

(iii) d (x, y) = d (y, x)

For if d (x, y) = 0. Then d (y, x) = 0 so that d (x, y) = d (y, x); and if d (x, y) = 1, then d (y, x) = 1 so that d (x, y) = d (y, x)

$\therefore$ In either case d (y, x) = d (x, y)

(iv) To prove d (x, y) + d (y, z) $\geq$ d (x, z).

If x = z, then d (x, z) = 0, so that

d (x, y) + d (y, z) $\geq$ 0 = d (x, z).

If x $\neq$ z, then d (x, z) = 1 and if y is any real number, then y is at least different from one of x and z ad hence at least one of d (y, x) and d(y, z) be equal to one. Consequently,

d (x, z) $\leq$ d (x, y) + d (y, z)

The above facts prove that d is a metric on R.

Example 26: *Prove that the function d : $R^2 \times R^2 \to R$ s.t.*

$$d(x, y) = [(x_1 - y_1)^2 + (x_2 - y_2)^2]^{1/2},$$

where $x = (x_{1,} x_2) = (y_2 - y_2) \in R^2$

Is a metric on R.

[This is called unusual metric on R^2]

Solution: Let $x = (x_{1,} x_2)$, $y = (y_{1,} y_2)$, $z = (z_1, z_2)$ be arbitrary elements of R^2

(i) d (x, y) $\geq$ 0.

For $(x_1 - y_1)^2 \geq 0$, $(x_2 - y_2)^2 \geq 0$ and hence

$$[(x_1 - y_1)^2 + (x_2 - y_2)^2]^{1/2} \geq 0$$

(ii) d (x, y) = 0 iff x = y

d (x, y) = 0

$$\Leftrightarrow (x_1 - y_1)^2 + (x_2 - y_2)^2 = 0$$

$$\Leftrightarrow x_1 - y_1 = 0,\ x_2 - y_2 = 0$$

$$\Leftrightarrow x_1 = y_{1,}\ x_2 = y_2$$

$\Leftrightarrow (x_{1,} x_2) = (y_{1,} y_2)$

$\Leftrightarrow x = y$

(iii) $d(x, y) = d(y, x)$.

For $[(x_1 - y)^2 + (x_2 - y_2)^2]^{1/2} = [(y_1 - x_1)^2 + (y_2 - x_2)^2]^{1/2}$

(iv) $d(x, y) + d(y, z) \geq d(x, z)$

For $d(x, y) + d(y, z) \geq [(x_1 - y_1)^2 + (x_2 - y_2)^2]^{1/2}$
$+ [(y_1 - z_1)^2 + (y_2 - z_2)^2]^{1/2} \geq [(x_1 - y_1 + y_1 - z_1)^2$
$+ (x_2 - y_{2+} y_2 - z_2)^2]^{1/2}$ be Minkowski inequality.

$= [(x_1 - z_1)^2 + (x_2 - z_2)^2]^{1/2} = d(x, z)$

$\therefore d(x, y) + d(y, z) \geq d(x, z)$.

Example 27: *Prove that the mapping $d : R^2 \times R^2 \to R$*

defined by $\quad d(x, y) = |x_1 - y_1| + |x_2 - y_2|$,

where $\quad x = (x_1, x_2),\ y = (y_1, y_2) \in R^2$.

is a metric on R^2.

Solution: Let $x = (x_1, x_2)$, $y = (y_1, y_2)$, $z = (z_1, z_2)$ be arbitrary elements of R^2. Also let

$$d(x, y) = |x_1 - y_1| + |x_2 - y_2|.$$

(i) $d(x, y) \geq 0$.

Since modulus of any quantity is always non-negative.

(ii) $d(x, y) = 0$ iff $x = y$.

For $d(x, y) = 0 \Leftrightarrow |x_1 - y_1| + |x_2 - y_2| = 0$

$\Leftrightarrow |x_1 - y_1| = 0 = |x_2 - y_2|$

$\Leftrightarrow x_1 = y_1, x_2 = y_2$

$\Leftrightarrow (x_{1,}x_2) = (y_{1,} y_2)$

$\Leftrightarrow x = y$.

(iii) $d(x, y) = d(y, x)$.

For $|x_1 - y_1| + |x_2 - y_2| = |y_1 - x_1| + |y_2 - x_2|$.

(iv) $d(x, y) + d(y, z) \geq d(x, z)$.

For $d(x, z) = |x_1 - z_1| + |x_2 - z_2|$

$= |x_1 - y_1 + y_1 - z_1| + |x_2 - y_2 + y_2 - z_2|$

$\leq |x_1 - y_1| + |y_1 - z_1| + |x_2 - y_2| + |y_2 - z_2|$

$= d(x, y) + d(y, z)$.

$\therefore \quad d(x, z) \leq d(x, y) + d(y, z)$.

Therefore d is a metric on R^2.

Example 28: *Show that the map* $d : R^2 \times R^2 \rightarrow R$

define by $d(x, y) = max. \{|x_1 - y_1|, |x_2 - y_2|\}$.

where $x = (x_1, x_2), y = (y_1, y_2) \in R_2$

is a metric on R^2.

Solution: Let $x = (x_1, x_2)$, $y = (y_1, y_2)$, $z = (z_1, z_2)$ be arbitrary elements or R^2. Also let

$d(x, y) = max \{|x_1 - y_1|, |x_2 - y_2|\}$

(i) $d(x, y) \geq 0$.

Since modulus of any quantity ;is non-negative.

(ii) $d(x, y) = 0$ iff $x = y$.

For $d(x, y) = 0 \Leftrightarrow max.\{|x_1 - y_1|, |x_2 - y_2|\} = 0$

$\Leftrightarrow |x_1 - y_1| = 0 = |x_2 - y_2|$

$\Leftrightarrow x_1 = y_1, |x_2 = y_2$

$\Leftrightarrow (x_1, x_2) = y_2, y_2)$

$\Leftrightarrow x = y$

(iii) $d(x, y) = d(y, x)$.

For $|a - b| = |b - a|$ and so

$max.\{|x_1 - y_1|, |x_2 - y_2|\} = max \{|y_1 - x_1|, |y_2 - x_2|\}$

(iv) $d(x, y) + d(y, z) \geq d(x, z)$

For $d(x, y) = max \{|x_1 - z_1|, |x_2 - z_2|\}$

$= max. \{|x_1 - y_1 + y_1 - z_1|, |x_2 - y_2 + y_2 - z_2|\}$

$\leq max. \{|x_1 - y_1| + |y_1 - z_1|, |x_2 - y_2| + |y_2 - z_2|\}$

$\leq max. \{|x_1 - y_1|, |x_2 - y_2|\} + max. \{|y_1 - z_1|, |y_2 - z_2|\}$.

$= d(x, y) + d(y, z)$.

The above facts lead to the conclusion that d is a metric on R^2.

Example 29: *Let X be a non-empty set and define d by*

$$d(x, y) = \begin{cases} 0 \text{ if } x = y \\ 1 \text{ if } x \neq y \end{cases}$$

Prove that d is metric on X.

Solution: Let $x, y, z \in X$ be arbitrary.

Then

(1) d (x, y) ≥ 0

This follows from the definition of d.

(2) d (x, y) = 0 iff x = y, by def. of d.

(3) d (x, y) = d (y, x) as value of d is free from order or x and y.

(4) d (x, y) + d (y, z) ≥ d (x, z)

For if x ≠ y ≠ z, then

$$d(x, y) + d(y, z) = 1 + 1 = 2 > 1 = d(x, z)$$

or, $$d(x, y) + d(y, z) > d(x, z) \quad \text{...(a)}$$

Again if x = y ≠ z, then

$$d(x, y) + d(y, z) = 0 + 1 = d = (x, z)$$

or, $$d(x, y) + d(y, z) = d(x, z) \quad \text{...(b)}$$

Combining (a) and (b), we get result (4).

∴ d is a metric on X.

Example 30: *Given two different metrics for the set R of all real numbers.*

Solution: Consider the map

$$d : R \times R \to R \text{ s. t.}$$

$$d(x,y) = \begin{cases} 0 \text{ if } x=y \\ 1 \text{ if } x \neq y \end{cases}$$

where (x, y) ∈ R^2.

Then d is metric on R.

Next consider the map

$$d' : R \times R \to R$$

$$d'(x, y) |x - y|.$$

where x, y ∈ R

Then d' is metric on R.

Thus d and d' are two different metrics on R.

Example 31: *If A is any non-empty subset of a metric space (X, d), prove that*

$$|d(x, A) - d(y, A)| \leq d(x, y)$$

for any points x, y ∈ X.

Solution: By def. of metric,

$$d(x, a) \leq d(x, y) + d(y, a)$$

and $d(y, a) \leq d(y, x) + d(x, a)$

Taking infimum over $a \in A$, we obtain

$$d(x, A) \leq d(x, y) + d(y, A)$$

and $$d(y, A) \leq d(y, x) + d(x, A)$$

i.e., $$d(x, A) - d(y, A) \leq d(x, y) \quad ...(1)$$

$$d(y, A) - d(x, A) \leq d(x, y)$$

The last gives

$$d(x, A) - d(y, A) \geq -d(x, y)$$

From (1) and (2),

$$-d(x, y) \leq d(x, A) - d(y, A) \leq d(x, y)$$

This $\Rightarrow |d(x, A) - d(y, A)| \leq d(x, y)$.

Example 32: *Let A, B be non-empty subsets of a metric space (X, d). Then show that*

$$d(A \cap B) \leq d(A) + d(B) + d(A, B)$$

Solution: Let $x, y \in A \cap B$ be arbitrary. Then there arise four cases.

(i) $x, y \in A$

This $\Rightarrow d(x, y) \leq d(A)$

For $d(A) = \sup \{d(x, y) : x, y \in A\}$

(ii) $x, y \in B$.

This $\Rightarrow d(x, y) \leq d(B)$

(iii) $x \in A$ and $y \in B$

(iv) $x \in B$ and $y \in A$

In case (iii), we have

$$d(x, y) \leq d(x, a) + d(a, y), \quad a \in A$$
$$\leq d(x, a) + d(a, b) + d(b, y), \; b \in B$$

$\therefore$ $$d(x, y) \leq d(x, a) + d(a, b) + d(b, y).$$

Similarly in case (iv), we have

$$d(x, y) \leq d(x, b) + d(b, a) + d(a, y)$$

This $\Rightarrow$

$$d(A \cup B) \leq d(B) + d(B, A) + d(A)$$

or $$d(A \cup B) \leq d(A) + d(B) + d(A, B).$$

Example 33: *Describe open spheres for the discrete metric on a space X.*

Solution: Let (X,d) be a discrete metric space so that d is given by

$$d(x,y) = \begin{cases} 0 \text{ if } x = y \\ 1 \text{ if } x \neq y \end{cases}$$

Let $x_0 \in X$ and $r > 0$ be any real number.

$$S_r(x_0) = \{x \in X : (x, x_0) < r\}.$$

Case (i) : When $r > 1$.

By definition metric, any $x \in X$, $x \neq x_0$

$$\Rightarrow d(x, x_0) = 1 < r$$

$$\Rightarrow d(x, x_0) = < r$$

$$\Rightarrow x \in S_r(x_0)$$

Also $x_0 \in S_r(x_0)$

$\because$ $$d(x_0, x_0) = 0 < r.$$

Combining these two results, we find that

any $$x \in X \Rightarrow x \in S_r(x_0)$$

This $$\Rightarrow X = S_r(x)$$

Case (ii) : when $r \leq 1$.

any $x \in X$, $x \neq x_0 \Rightarrow d(x, x_0) = 1 < r$

$$\Rightarrow d(x, x_0) < r.$$

$$\Rightarrow x \notin S_r(x_0)$$

$$x_0 \in S_r(x_0)$$

Thus, any $x \in X$, $x \neq x_0 \Rightarrow x \notin S_r(x_0)$

and $$x_0 \in S_r(x_0).$$

From this it follows that $S_r(x_0) = \{x_0\}$.

Finally, $$S_r(x_o) = \begin{cases} X & \text{if } r . 1. \\ \{x_0 & \text{if } r \leq 1.\} \end{cases}.$$

Example 34: *Describe open spheres for the usual* metric on R, *i.e.*, open spheres on the real line are precisely the bounded open intervals.

Solution: Let d be the usual metric on R; then

$$d(x, y) = |x - y| \ \forall \ x, y \in R$$

Let $x_0 \in R$ and $r > 0$ be any real number.

$$\begin{aligned} S_r(x_0) &= \{x \in R : d(x, x_0) < r\} \\ &= \{x \in R : |x - x_0| < r\} \\ &= \{x \in R : -r < x - x_0 < r\} \\ &= \{x \in R : x_0 - r < x < x_0 + r\} \\ &= (x_0 - r, x_0 + r). \end{aligned}$$

This proves that the open spheres on the real line R are open intervals.

Example 35: *Describe open spheres for the usual metric on R^2.*

Solution: Let d be the usual metric on R^2 so that d is given by

$$d(x, y) = [(x_1 - x_1)^2 + (y_1 - y_2)^2]^{1/2}$$

where $x = (x_1, y_1) \in R^2, y = (x_2, y_2) \in R^2$.

Let $z_0 = (x_0, y_0) \in R^2$ and $r > 0$ be any real number

$$\begin{aligned} S_r(z_0) &= \{(x, y) \in R^2 : z = (x, y), d(z, z_0) < r\} \\ &= \{(x, y) \in R^2 : z = \sqrt{[(x - x_0)^2 + (y - y_0)^2]} < r\} \\ &= \{(x, y) \in R^2 : (x - x_0)^2 + (y - y_0)^2 < r^2\} \end{aligned}$$

This proves that open sphere $S_r(z_0)$ is the interior of the circle $(x - x_0)^2 + (y - y_0)^2 = r^2$, where $z_0 = (x_0, y_0)$.

Example 36: *Describe open sphere of unit radius having the centre at (0, 0) for each of the following metric* ..: *R.*

(i) $d_1(z_1, z_2) = |x - x_2| + |y_1 - y_2|$

(ii) $d_2(z_1, z_2) = \max. \{|x_1 - x_2|, |y_1 - y_2|\}$

where $z_1 = (x_1, y_1), z_2 = (x_2, y_2) \in R^2$.

Solution: Let $z_0 = (0, 0), z = (x, y) \in R^2$.

To determine $S_1(z_0)$.

Step (i)

$$\begin{aligned} S_1(z_0) &= \{(x, y) \in R^2 : d_1(z, z_0) < 1\} \\ &= \{(x, y) \in R^2 : |x - 0| + |y - 0| < 1\} \\ &= \{(x, y) \in R^2 : |x| + |y| < 1\} \end{aligned}$$

This shows that the open sphere $S_1(z_0)$ is the interior of the square bounded by the lines

$$x + y = 1, -x - y = 1, -x + y = 1, x - y = 1.$$

Step (ii) $S_1(z_0) = \{(x, y) \in R^2 : d_1 (z, z_0) < 1\}$

$= \{(x, y) \in R^2 : \max. \{|x - 0|, |y - 0|\} < 1\}$

$= \{(x, y) \in R^2 : \max. \{|x|, |y|\} < 1\}$

This proves that the open sphere $S_1(z_0)$ is the interior of the square bounded by the lines

$x = 1, - x = 1, y = 1. - y = 1.$

Example 37: *Contor's set contains exactly as many points as the interval [0, 1] .*

Hint : To prove $|E| = |[0,1]|$.

Also $|[0,1]| = C$

E is a closed set.

Since an infinite disjoint union of open sets is open, Hence the family of removed open intervals form an open set.

$E = [0,1]$ – family of removed open intervals.

$= [0,1]$ – open set = closed set.

For complement of open set is a closed set.

Example 38: *Given an example of two closed subset A and B of the real line r such that d (A, B) = 0, but A/B ∅ .*

Solution: Let d be the usual metric on R. Write

$N_1 = \{ 2, 3, 4, 5 \ldots \} = \{n + 1 : n \in N\}$

$A = \{2, 3, 4, 5 \ldots\},$

$$B = \left\{2\frac{1}{2}, 3\frac{1}{3}, 4\frac{1}{4}, 5\frac{1}{5}\right\}$$

Then A and B are expressible as

$A = \{n : n \in N_1\},$

$$B = \left\{n + \frac{1}{n} : n \in N_1\right\}$$

If $n \in A$, then $n + 1/n \in B$ and the distance between $n + 1/n$ and n is $1/n$, which $\to 0$ as $n \to \infty$.

$d (A, B) = \inf \{d (x, y) : x \in A, y \in B\}$

$= \inf \{| x - y| : x \in A, y \in B\}$

$= \inf \{1/n : n \in A, n + 1/n \in B\} = 0.$

Thus, d (A, B) = 0. Also $A \cap B = \varnothing$.

Therefore A and B are the required sets.

Example 39: *Every set in a discrete metric space is open.*

Solution: Let (X,d) be a discrete metric space. Let x, y $\in$ X be arbitrary.

By def. of discrete metric, d (x, y) = $\begin{cases} 1 \text{ if } x \neq y \\ 0 \text{ if } x = y \end{cases}$.

Let r be any positive real number s.t. $r \leq 1$.

Then $$\begin{aligned} S_r(x) &= \{y \in X : d(y, x) < r \leq 1\} \\ &= \{y \in X : d(y, x) < 1\} \\ &= \{y \in X : d(y, x) = 0, \text{ by def. of d.} \\ &= \{y \in X : y = x\} = \{0\}. \end{aligned}$$

or, $$S_{r(x)} = \{0\}.$$

But every open sphere is an open set.

$\therefore$ {x} is an open set in X $\forall$ x $\in$ X.

If A = $\{x_1, x_2, ..., x_n\}$ = finite set $\subset$ X, then

$A = \bigcup_{r=1}^{n} \{x_r\}$ = finite union of open sets.

= open set.

Hence every finite sub set of X is open set. ...(1)

If $$B = \{x_1, x_2, x_3, ...\} \subset X, \text{ then}$$

B is an infinite subset of X.

Now $$B = \bigcup_{r=1}^{\infty} \{x_r\}$$

= Arbitrary union of open sets

= open set, by Theorem 8.

$\therefore$ B is an open set, ...(2)

From (1) and (2), it follows that every subset (finite or infinite) is an open set in X.

Corollary : In a discrete metric space, every subset of X is closed.

Solution: We know that G is open

$\Leftrightarrow$ X – G = G' is closed set.

Let A be a subset of a discrete metric space (X, d), then A is open.

Also A' = X – A is again sub set of X and so A' is open. Consequently A is closed.

Thus every subset of X is open as well as closed.

Example 40 : *A finite set in any metric space has no limit point.*

Solution: Let A be a finite subset of a metric space (X, d). We know that "$x \in X$ is a limit point of any set B if every open sphere $S_{r(x)}$ contains an infinite number of points of B other than x."

This condition can not be satisfied here as A is finite set.

Hence A has no limit point.

EXERCISES

1. Find closure of integers, rationals, Contor's set on the real line.
2. $d(z_1, z_2) = |z_1 - z_2|$ is a metric on the set C of complex numbers.
3. Show that Cantor set is nowhere dense.
4. Let $S_r(x)$ be an open sphere in a metric space (X, d) and $A \subset X$ with diameter < r which intersects $S_r(x)$. Prove that
$$A \subset S_{2r}(x).$$
5. Let a be a subset of a metric space (X, d). Then prove that "a is dense $\Leftrightarrow$ the only closed super set of A is X $\Leftrightarrow$ the only open set disjoint from A is ϕ $\Leftrightarrow$ A intersects every sphere.
6. Prove that Cauchy sequence is convergent $\Leftrightarrow$ it has a convergent subsequence.
7. Prove that the set of all real continuous functions defined on [0, 1] is a metric space if we define
$$d(f, g) = \sup_{0 < t \le 1} |f(t) - g(t)|$$
as the distance between the functions f and g.

   How is the metric convergence related to uniform convergence.
8. Let A and B be two subsets of a metric space X.

   Prove that :

   (i) $\text{Int } A \cup \text{Int }(B) \subseteq \text{Int }(A \cup B)$

   (ii) $\text{Int } A \cap \text{Int }(B) = \text{Int }(A \cup B)$.

   Give an example on the real line to show that inclusion in (i) above can be strict.

10. Define convergence of a sequence in a metric space (X, d). Prove that, for $A \subset X$, $x_0 \in X$, the point x_0 is a limit point of the set A iff there is a sequence of distinct points of A converging to x_0.

11. State and prove Minkowski's inequality.

12. (a) Define a space and give one example.

    Let ρ be a metric on X. Show that ρ defined by

    $$\sigma(x, y) = \frac{\rho(x,y)}{1+\rho(x,y)} \text{ is a metric on X.}$$

    (b) Define an open ball; and an open set in a metric space. Prove that a finite intersection of open sets is open. Give an example to show that an infinite intersection of open sets need not be open.

13. Define pointwise convergence and uniform convergence of a sequence of functions on a metric space X into a metric space y.

14. (a) Define a metric space. Prove that the set of all bounded continuous real functions defined on the closed unit interval has a metric defined by

    $$d(f, g) = \sup \{|f(x) - g(x) \in [0, 1]\}$$

    (b) When is a subset of a metric space called open ?

    Prove that in a metric space :

    (i) any union of open sets is open.

    (ii) any finite intersection of open sets is open.

15. (a) Prove that in any metric space, each closed sphere is a closed set.

    (b) State and prove Cantor's intersection theorem.

16. (a) Define a metric on a set X. Prove that on the complex plan function d defined by

    $$d(z_1, z_2) = |z_1 - z_2|$$

    is a metric

    (b) Define an open set in a metric space (X, d) and prove that

    (i) union of open sets in (X, d) is open.

    (c) Let $S_r(x)$ be an open sphere in a metric space X whose centre is x and radius r. If A is a subset of X with diameter less than r which intersects then prove that $A \subset S_{2r}(x)$.

17. State Cauchy's inequality and use it to prove Minkowski's inequality.
18. If d is a metric on X, prove that ρ defined by

$$\rho(a, b) = \frac{d(a, b)}{1 + d(a, b)}$$

is also a metric on X.

19. Define the concepts of a metric space and that of an open set in such a space.
20. Define the concept of a metric space. Show that subset F of a metric space X is closed ⇔ its complement F is open.
21. Let r be the set of all real numbers and let

$$d(x,y) = \frac{|x-y|}{1+|x-y|} \text{ for all } x, y \in R$$

Prove that d is a metric on R.

22. (a) Show that in a metric space every finite set is closed.

    (b) Show by means of an example that the union of an infinite collection of closed sets need not be closed.
23. Define the distance d (A, B) between two sets A and B in a metric space.
24. Let X be a metric space with metric d. Show that d_1, defined by $d_1(x, y) = d(x, y)/[1 + d(x, y)]$, is also a metric on X. Observe that X itself is a bounded set in the metric space (X, d_1).
25. Let X be a non-empty set, and let d be a real function of ordered pairs of elements of X which satisfies the following two conditions: $d(x, y) = 0 \Leftrightarrow x = y$, and $d(x, y) \le d(x, z) + d(y, z)$. Show that d is a metric on X.
26. Show that a non-empty subset A of a Banach space is bounded ⇔ there exists a real number K such that $\|f\| \le K$ for every x in A.
27. Construct a sequence of continuous functions defined on [0, 1] which converges pointwise but *not* uniformly to a continuous limit.
28. Construct a sequence of continuous functions defined on [0, 1] which converges pointwise to a discontinuous limit.
29. Let X and Y be metric spaces with metrics d_1 and d_2, and $\{f_n\}$ $\{f_n\}$ be a sequence of mappings of X into Y which converges pointwise to a mapping f of X into Y, in the sense that $f_n(x) \to f(x)$ for each

x in X. Define what ought to be meant by the statement that f_n converges uniformly to f and prove that under this assumption f is continuous if each f_n is continuous.

30. Let X and Y be metric spaces and f a mapping of X into Y. If f is a constant mapping, show that f is continuous. Use this to show that a continuous mapping need not have the property that the image of every open set is open.

31. Let X be a metric space with metric d, and let x_0 be a fixed point in X. Show that the real function f_{x_0} defined on X by $f_{x_0}(x) = d(x, x_0)$ is continuous. Is it uniformly continuous?

32. Let X and Y be metric spaces and A a non-empty subset of X. If f and g are continuous mappings of X into Y such that f(x) = g(x) for every x in A, show that f(x) = g(x) for every x in $\overline{A}$.

33. Let X and Y be metric spaces and f a mapping of X into Y. Show that f is continuous $\Rightarrow$ $f^{-1}(F)$ is closed in X whenever F is closed in Y $\Leftrightarrow$ $f(\overline{A}) \subseteq \overline{f(A)}$ for every subset A of X.

34. Show that any mapping of the metric space any other metric space is continuous.

35. Consider the real function f defined on the real line R by $f(x) = x^2$. If b is a given positive real number, show that the restriction of f to the closed interval [0, b] is uniformly continuous by starting with an $\epsilon > 0$ and exhibiting a $\delta > 0$ which satisfies the requirement of the definition.

36. Determine which of the following functions are uniformly continuous on the open unit interval (0, 1) : $1/(1 - x)$; $1/(2-x)$; sin x; sin (1/x); $x^{1/2}$, x^3. Which are uniformly continuous on the open interval $(0, +\infty)$?

37. In the proof of Theorem D we used the following fact: the image of a Cauchy sequence under a uniformly continuous mapping is again a Cauchy sequence. Give the details of the proof.

38. Let f be a continuous real function defined on R which satisfies the functional equation f(x + y) = f(x) + f(y). Show that this function must have the form f(x) = mx for some real number m.

    (Hint: the subspace of rational numbers is dense in the metric space R.)

39. Let X be a metric space, proving the following statements:

(a) any point and disjoint closed set in X can be separated by open sets, in the sense that if a: is a point and F a closed set which does not contain x, then there exists a disjoint pair of open sets G_1 and G_2 such that $x \varepsilon G_1$ and $F \subseteq G_2$;

(b) any disjoint pair of closed sets in X can be separated by open sets, in the sense that if Fi and Fi are disjoint closed sets, then there exists a disjoint pair of open sets Gi and Gi such that $F_1 \subseteq G_1$ and $F_2 \subseteq G_2$.

40. Let X be a metric space, and let A be a subset of X. If a: is a limit point of A, show that each open sphere centered on x contains an infinite number of distinct points of A. Use this result to show that a finite subset of X is closed.

41. Show that a subset of a metric space is bounded <=> it is non-empty and is contained in some closed sphere.

42. Give an example of an infinite class of closed sets whose union is not closed. Give an example of a set which :

(a) is both open and closed;

(b) is neither open nor closed;

(c) contains a point which is not a limit point of the set; and

(d) contains no point which is not a limit point of the set.

43. Describe the interior of the Cantor set.

44. Prove the statements made in the text about closures.

45. Let X be a metric space and A a subset of X. Prove the following facts:

(a) $\overline{A}$ = Int (A');

(b) $\overline{A}$ = {x:d(x, A) = 0}.

46. Describe the closure of each of the following subsets of the real line: the integers; the rationals; the Cantor set; (0, + ∞); (–1, 0) ∪ (0, 1). Do the same for each of the following subsets of the complex plane: {z:|z| is rational}; {z:l/R(z) is an integer}; {z:\z\<1 and I(z) < 0}.

47. Let X be a metric space, let a; be a point of X, and let r be a positive real number. One is inclined to believe that the closure of $S_r(x)$ must equal $S_r[x]$. Give an example to show that this is not necessarily true.

48. Let X be a metric space, and let G be an open set in X. Prove that G is disjoint from a set A ⇔ G is disjoint from A.

49. Prove the facts about boundaries stated in the text.

50. Describe the boundary of each of the following subsets of the real line: the integers; the rationals; [0, 1]; (0, 1). Do the same for each of the following subsets of the complex plane: $[z:|z| < I\}$; $\{z:|z| \leq l\}$; $[z:I(z)>0\}$.

51. Let X be a metric space and A a subset of X. A is said to be dense (or *everywhere dense*) if $\overline{A} = X$. Prove that A is dense $\Leftrightarrow$ the only closed superset of A is X $\Leftrightarrow$ the only open set disjoint from A is ϕ $\Leftrightarrow$ A intersects every non-empty open set $\Leftrightarrow$ A intersects every open sphere.

52. Let X be a metric space. If $\{x_n\}$ and $\{y_n\}$ are sequences in X such that $x_n \to x$ and $y_n \to y$, show that $d(x_n, y_n) \to d(x, y)$.

53. Show that a Cauchy sequence is convergent $\Leftrightarrow$ it has a convergent subsequence.

54. If $X = X_1 \times X_2 \times \ldots \times X_n$ is the product, and if each of the coordinate spaces $X_1, X_2, \ldots, X_n$ is complete, show that X is complete with respect to each of the metrics d and $\overline{d}$ defined in that problem.

55. Let X be any non-empty set. The set of all bounded real functions, defined on X is a metric space with respect to the metric induced by the norm defined in that problem. Show that this metric space is complete. (Hint: if $\{f_n\}$ is a Cauchy sequence, then $\{f_n(x)\}$ is a Cauchy sequence of real numbers for each point x in X.)

56. Give an example to show that the set F in Cantor's intersection theorem may be empty if the hypothesis $d(F_n) \to 0$ is dropped.

57. Show that a closed set is nowhere dense $\Leftrightarrow$ its complement is everywhere dense.

58. Show that the Cantor set is nowhere dense.

59. Let X be a metric space, and show that any two distinct points of X can be separated by open spheres in the following sense: if x and y are distinct points in X, then there exists a disjoint pair of open spheres each of which is centered on one of the points.

60. Let X be a metric space. If $\{x\}$ is a subset of X consisting of a single point, show that its complement $\{x\}'$ is open. More generally, show that A' is open if A is any finite subset of X.

61. Let X be a. metric space and $S_r(x)$ the open sphere in X with center x and radius r. Let A be a subset of X with diameter less than r which intersects $S_r(x)$. Prove that $A \subseteq S_{2r}(x)$.

62. Let X be a metric space. Show that every subset of X is open $\Leftrightarrow$ each subset of X which consists of a single point is open.

63. Let X be a metric space with metric d, and let d_1 be the metric defined. Show that the two metric spaces (X, d) and (X, d_1) have precisely the same open sets. (*Hint:* show that they have the same open spheres with one exception. What is this exception?)

64. If $X = X_1 \times X_2 \times ... \times X_n$ is the product, and if d and 3 are the metrics on X defined in that problem, show that the two metric spaces (X, d) and (X, d) have precisely the same open sets. Observe that in this case the spaces do not have the same open spheres.

65. Let Y be a subspace of a metric space X, and let A be a subset of the metric space Y. Show that A is open as a subset of Y $\Leftrightarrow$ it is the intersection with Y of a set which is open in X.